EXCAVATORS

EXCAVATORS

Peter N. Grimshaw, BSc, MSc, FRGS, FGS

BLANDFORD PRESS
POOLE · DORSET

TO PETER THOMAS AND HIS GRANNIE

First published in the U.K. 1985 by Blandford Press,
Link House, West Street, Poole, Dorset, BH15 1LL

Copyright © 1985 Peter N. Grimshaw

Distributed in the United States by
Sterling Publishing Co., Inc.,
2 Park Avenue, New York, N.Y. 10016

British Library Cataloguing in Publication Data

Grimshaw, Peter N.
  Excavators.
  1. Earthmoving machinery
  I. Title
  629.2'25   TA725

ISBN 0 7137 1335 6

Typeset by Keyspools Ltd, Golborne, Lancs
Printed in Italy
by Interlitho, Milan

BST
R

#20.00

Contents

Acknowledgements and Author's Note

The author is indebted to many organisations world-wide which have responded to requests for information and materials, and to individuals who have so willingly given of their time and expertise during the preparatory research for this book: only a few can be recorded. Marion Power Shovel Division of Dresser Industries Inc. generously sponsored an extensive study-tour of the USA and Canada in 1983: Nikon UK Ltd assisted with the loan of additional photographic equipment. Especial thanks for help is due to Jim Williams, Dwight Wilcox and Peter Gilewicz (Marion-USA) and Harry Webb (Marion-UK); Pierre Trullas (Poclain-France) and Brian Shaw (Poclain-UK); Ivan Jameson (NCB Opencast Executive-UK); Keith Haddock (Luscar Ltd-Canada) and Royston Bulley (UK). Keith Haddock's additional hospitality and encouragement have been greatly appreciated; and the help and forbearance throughout many interruptions to home life of my wife Joan are acknowledged with love and gratitude.

Most illustrative material has been provided by excavator manufacturers. The plates on pages 15, 23 (upper), 26, 27, 30, 41 (upper), 43, 46, 51 (upper), 58 (upper), 62, 71 (upper), 74, 83 (lower), 90, 98 (upper), 102, 123, 126 (lower), 128, 130, 134 (upper), 139, 142 (lower), 155 (upper), 156 (lower), 157 and 158 are by the author: the plate on p. 17 is reproduced by courtesy of Leicestershire Museums, Art Galleries and Records Service; the lower plate on p. 18 by Manchester Ship Canal Company; the plate on p. 53 by Novosti Press Agency (APN); the upper plate on p. 156 by the US National Park Service/Dwight D. Eisenhower Library.

In general, 'current' data refers to the situation appertaining in 1983: in cases when repeated attempts failed to elicit such information, the most recent available has been substituted. A limited amount of information on significant developments during 1984 has also been incorporated. Although every effort has been made to ensure accuracy by checking the validity and comparability of data, neither the author nor the publishers can accept responsibility for any errors or omissions. The author would, however, welcome comment and additional information.

Statistics have been rounded off for clarity where necessary. It must also be borne in mind that manufacturers not only use different standards for specifications, but also regularly produce new models and versions of machines, while reserving the right to amend specifications without prior notice. Metric units are given throughout; but as the imperial system is widely employed by excavator manufacturers and users these units are given as well for certain features.

P.N.G.
July 1984

CHAPTER 1
Introduction to Earthmoving

Moving the earth is a fundamental human activity. Children defy advancing tides at the seaside with buckets and spades to make transient castles and canals, hills and hollows: the old relive their creative youth in treasured gardens by turning over and re-moulding small portions of the earth's surface to make them both attractive and productive.

The act of digging must have always been of great significance. In ancient times building materials were obtained for constructing shelters; primitive settlements were protected by moats and upstanding earthworks; waterways were dug to irrigate or drain the land; underground chambers and mounds were made for the burial and commemoration of the dead; and later minerals and fossil fuels were extracted from shallow workings, and major routeways built, to provide the basis of industry and trade. Early examples of earthmoving were often no mean feats. King Cheops' Great Pyramid at Giza, Egypt, was built some 4,600 years ago and involved moving around 1.2 million m³ (1.6 million cu yd) of the earth's surface; the Great Wall of China, over 2,000 years old and stretching for some 3,220 km (2,000 miles) required no less than 55 million m³ (72 million cu yd) of material for its construction.

In more recent times the difficulties inherent in underwater excavation resulted first in floating

Mechanical excavators were first devised for dredging: this single-bucket machine was illustrated in a book by Austrian Bishop Verantius in 1591.

devices to dredge rivers, dig canals and construct harbours. In 1513 the indomitable genius Leonardo da Vinci produced a design for an underwater digging machine; there exists a drawing of a bucket chain dredger used in Belgium on the Rupel-Scheldt canal in 1560; and an illustration in a book by the Austrian Bishop Verantius, dated 1591, appears to show a pontoon-mounted grab dredger operated by hand winches and a treadmill. Predictably, therefore, the initial application of steam, the new source of power developed in the eighteenth century, to earthmoving was to water-based equipment. The first steam excavating machine was apparently designed by Grimshaw of Sunderland, Tyne and Wear, in 1796 when he arranged for Boulton & Watt to build and install a 3 kW (4 HP) engine in a scow – a flat-bottomed boat – to work dredging machinery.

For centuries little progress had been made in the development of dry-land earthmoving methods: spades, picks, rakes and trowels used to dig, and sledges, wheelbarrows and containers suspended from yokes employed to carry, remained fundamentally the same. Animal power supplemented human labour. Despite, however, the advent of steam power and the development of primitive machines for underwater excavation, these dry-land traditional methods of earthmoving were employed well into the industrial revolution of the nineteenth century. Labour then was both cheap and plentiful, certainly in many countries for a long time, and there was little incentive to apply steam power to dry-land projects. The bulk of the UK's navigation canals was created, in the late eighteenth century and early nineteenth century, solely by hand labour.

Similarly, the great majority of the cuttings, embankments and tunnels required for that country's railway system – on which construction commenced in earnest in the 1820–30 period and which was largely completed by the end of the century – were made by navvies: labourers so called since their predecessors dug the UK's network of navigation canals.

Nevertheless, in North America the situation was somewhat different. Here a vast continent, being settled over large areas for the first time, required extensive earthmoving projects; particularly those associated with railway construc-

tion opening up 'the West'. The scale of these operations, the speed required and a relative scarcity of labour were factors which led to early experiments with powered dry-land earthmoving machines. William Otis responded by providing the first practical excavators in the 1830s and 40s, although even in the USA it was to be another twenty or thirty years before the incipient excavator manufacturing industry began to gain substantial momentum.

Excavators based on the principles established by William Otis – with single buckets and with at first limited slew but, after 1884, developing into full 360° slew machines – were to become the prime earthmoving machines. The smaller types include the most versatile base machines of all construction and escavation equipment; the larger more specialised ones are amongst the biggest self-propelled land vehicles on earth. These then, with bucket capacities from 0.08 to 168 m³ ($\frac{1}{10}$–220 cu yd) are the subject matter of this book.

Nevertheless, excavators have not been developed, or used, in isolation from other forms of excavating machines. Prior to the application of powered machines to earthmoving in the USA, horse-drawn blades and scoops were often used in conjunction with wooden carts pulled by mules. Later, small metal scoops which were pulled by horses and guided by an operator were tipped to fill, dragged, and tipped again to dump. Primitive 'scrapers' of this type, including the Fresno patented in 1882 by James Porteous, a blacksmith in the town of Fresno, California, were used throughout the nineteenth century for light excavation tasks, eventually motive power was provided for larger wheeled models by steam traction engines. It was R. G. Ke Tourneau, originally also from Stockton, California, who developed the scraper into its modern form: a machine which performs three basic functions – loading, hauling and dumping.

Although it was in England where the first caterpillar tracks (crawlers) were developed – Richard Edgeworth of Hare Hatch, Berkshire, patented a system as early as 1770 – it was yet again in California at the beginning of this century where practical developments of major significance took place. Benjamin Holt of Stockton and Daniel Best of San Leandro laid the foundations of what was to become the Caterpillar Tractor Company. However, in the UK David Roberts of Richard Hornsby & Sons of Grantham, Lincolnshire, played an equally important design role in the formative years. Crawler tractors soon logically developed into earthmoving bulldozers and loading shovels as well as haulage units for scrapers: now Caterpillar's D10 bulldozer is of no less than 522 kW (700 HP).

Wheel loaders and bulldozers were evolved from agricultural tractors in the 1920–30 period, the name of R. G. Le Tourneau again being in evidence amongst others such as Frank G. Hough. The grader, like the scraper, has a long ancestry. Horse-drawn wagons with adjustable blades were used for tasks such as levelling roads early in the nineteenth century but in 1885 J. D. Adams, a road inspector from Indianapolis, Indiana, invented a 'Little Wonder' unit whose two wheels could be made to lean against the sideways thrust of an angled blade. The Adams company, and also later the Russell company, pioneered subsequent developments in the USA.

Backhoe loaders also evolved from agricultural tractors following advances in hydraulics technology during and soon after World War 2; Case in the USA with JCB and Whitlock in the UK were early entrants into this field of earthmoving. Trenchers for land-drainage and pipe-laying work first appeared in the USA around 1900 with manufacturers including the Cleveland, Barber-Greene and Buckeye companies. The similar, but larger, bucket chain and bucket wheel excavators originated in Europe. In Germany Lübecker of Lübeck, and in France Boulet of Paris, were makers of bucket chain machines at the end of the nineteenth century. The bucket wheel excavator has been developed to enormous proportions in East and West Germany; by companies such as O & K, Krupp & Buckau-Wolf in West Germany with individual machine weights approaching 14,000 t and daily outputs of 240,000 m³ (314,000 cu yd) of material. Large models of this type of machine have also been made in the USSR and USA.

All of these other types of earthmoving machines – scrapers, crawler and wheel loading shovels and bulldozers, graders, backhoe loaders, trenchers, bucket chain and bucket wheel excavators, have limitations to their usage: none can compete with excavators in range of applications through their versatility and immense available power. Excavators dig the foundations for buildings and excavate the raw materials for their construction including clay, limestone, sand and gravel; create transportation features such as canals, roads, railways and harbours; undertake earthmoving tasks for utilities including reservoir construction, trenches for pipelines and irrigation and flood control work; extract the bulk of the world's minerals exploited by surface mining, for example, copper, iron ore and bauxite; play a large and increasing role in coal production obtained by surface mining; and provide phosphate for fertilizer production.

It is an appropriate time to undertake a review of these machines and the industry which produces them; a hundred and fifty years have elapsed since the first excavator was made and a hundred since the initial full slew machine was devised. This book is an affirmation of the value of the machines themselves: impressive machines which moved the British author J. B. Priestley after a visit to Ruston-Bucyrus' factory in 1933 when compiling his *English Journey*, to write of 'these giant excavators, which sink their teeth into the earth and bite away tons and tons of it in a few minutes'; machines which emulated the movements of human limbs long before electronic robots were conceived; machines which, despite man's apparent inability consistently to match available resources in the developed world to ever-increasing latent demand elsewhere, are nevertheless still required in great quantities to relieve poverty and provide acceptable standards of living; and machines which, in the opinion of one German professor, may well be of therapeutic value in the year 2000 as large-scale earthmoving will never be (at least for as far as can be adjudged) fully automated and will always require the active participation of man.

The book is, though, really a tribute to people – the countless numbers who, over a century and a half, have designed, manufactured and put to use excavators to ensure that 'Progress begins with Digging'.

CHAPTER 2
Early Development of Excavators

A contributor to the *Troy Daily Whig* newspaper of Troy, New York State, wrote in the early 1840s 'It is one of those rare inventions, where genius seems to have worked out for itself an independent field of action, and the benefits flowing from which are destined to be known and felt throughout the civilized world.' How right he – or she – was. The reference was to one of the first steam shovels, newly invented by William S. Otis, and then at work on the Schenectady and Troy Railroad where it 'continues to excite a lively interest amongst our citizens, thousands of whom have been prompted by curiosity to pay it a visit.'

At least two descendants of John Otis, who emigrated from England in 1631, and settled at Hingham, Massachusetts, were destined to make significant inventions, although history books only give due acknowledgement to the one associated with the creation of passenger lifts. William S. Otis, however, born in 1813 as the son of the postmaster at Pelham, Massachusetts, is no less worthy of fame. William's family moved in either 1828 or 1829 to Philadelphia, Pennsylvania, where in 1833 he joined the contracting firm of Carmichael & Fairbanks of which the senior partner, Daniel Carmichael, was his brother-in-law: by the following year he was himself a partner.

When Carmichael, Fairbanks & Otis had a contract to work on the construction of the Boston and Providence Railroad (later the New York, New Haven and Hartford) in 1835, William Otis made his temporary home at Canton, Massachusetts. Here, not only did he meet and marry a local girl, but also – with the help of a local man, Charles Howe French, recently out of apprenticeship to a millwright – he designed and built the prototype of the world's first successful, powered, dry-land excavator. This, built in a machine shop near the Massapoag Brook, was probably based on the quarry derrick principle and is believed to have had a mast supported by cables, a 'luffing-type' dipper arm and bucket hinged to a boom, a double drum hoisting engine, and boom swing achieved by men at the sides pulling on ropes. The machine was clearly not easily moved as the derrick and excavating equipment were not fixed to a readily mobile frame also carrying the engine.

This Canton-built prototype was later used by Charles French on construction of the Norwich and Worcester Railroad; and was possibly the same machine, incorporated into a dredge designed by Hosea T. Stock, which Charles French and Daniel Carmichael operated as contractors widening the Welland Canal, Ontario, in the 1840s and 1850s.

William Otis returned to Philadelphia in 1836, intent on building an improved version of his excavator himself with parts made by local machine shops. In the event he entrusted the complete erection of seven machines to Garret & Eastwick, builders of locomotives and stationary engines, with premises on Twelfth Street. The first Philadelphia-built Otis Shovel was produced in 1837 and by the time the second emerged the firm's name had become Eastwick & Harrison; Joseph Harrison Jr, the skilled foreman who is likely to have contributed much to perfecting mechanical details of the shovel's design, had bought out Garret's interest.

Although William Otis had filed his original patent application on 15 June 1836, a fire in the patent office destroyed this and a second application dated 27 October 1838 was allowed on 24 February 1839 and numbered 1089. In this restored patent application for a 'Crane-excavator for excavating and removing earth' the inventor did not 'make any claim to the mere using of the scraper by means of the crane, this having been done before ... I do claim ... the

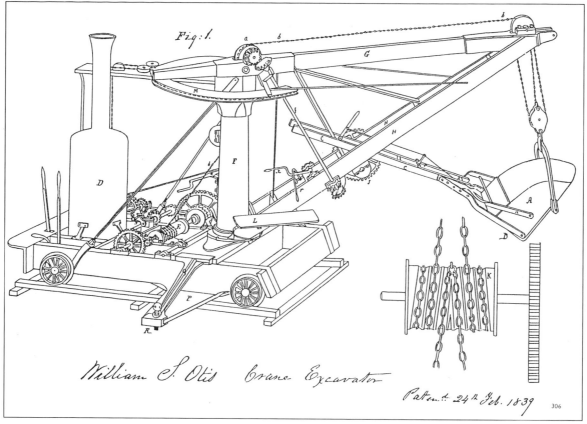

Fig: 1.

William S. Otis Crane Excavator

Patent 24th Feb. 1839

306

Otis' drawing which accompanied his restored patent granted in 1839.

application of power to force the scraper forward against a bank in the act of excavating and to withdraw it at pleasure ... also the general combination of the friction belt around the pulley upon the shaft (for carrying the scraper out and in as may be required) and the apparatus for tightening the same, or of allowing it to run loose ...'. The Otis shovel was, therefore, the first excavator to break and remove material by employing a single bucket with power thrust for adjusting the radius of the cut above, and to a limited extent below, ground level. It was self-propelled and exhibited the three basic digging features still present in excavators equipped as shovels: these are hoist, crowd (or racking) and slew (or swing).

Philadelphia-built Otis shovels had a wide

Shovels based on Otis' design were marketed throughout the second half of the nineteenth century.

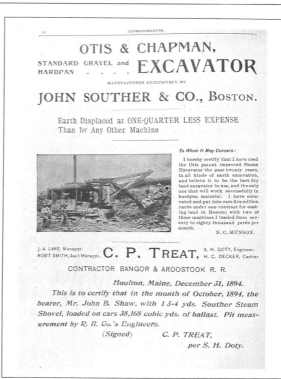

timber frame supported on cast iron travelling wheels which ran on temporary rails; the frame carried the boiler, engine and gears as well as all the excavating equipment and had projections with screw jacks for added stability when working. A triangular timber boom was supported by a pivoted, tubular, cast iron centre post with a semi-circular slewing ring attached to its top. Power for hoisting was taken by chain from a grooved drum, through the inside of the centre post to pulleys on the post and boom and then down to the bucket (scraper). A shaft with bevel wheels at each end carried power down from the post pulley to another shaft between the main lower braces of the boom which, with its appendages, was the key feature of the Otis shovel. A drum fitted to this shaft carried chains in reversed direction attached to opposite ends of the dipper arm: this device provided controlled thrust with the capability of absorbing shocks, to the arm and its bucket of around 0.95 m³ (1¼ cu yd) capacity. Up to 180° of slew was achieved by means of a chain from another drum taken round the slewing ring. At the rear of the machine, a vertical wood-burning boiler provided low-pressure steam to a single cylinder vertical engine, with approximately a bore of 23 cm (9 in) diameter and 30 cm (1 ft) stroke, fitted at its front; this resulted in units ranging from about 6 to 15 kW (8–20 HP). Contemporary accounts indicate that an Otis shovel could do the work of between 60 and 120 men: its estimated output was around 76 m³ (100 cu yd) of material per hour.

Of the seven Philadelphia-built shovels only two remained in North America. There are suggestions that the first work undertaken by one of these was on the Baltimore and Ohio Railroad in Maryland, but the first incontrovertible use of a powered excavator was in 1838 at Springfield, Massachusetts, when Carmichael, Fairbanks & Otis were building a link of the Western Railroad (later Boston and Albany) to Worcester. A local newspaper here referred to the Otis shovel as 'a specimen of what the Irishmen call "digging by stame".' The writer obviously felt Isaiah's prophecy – immortalised by Handel in the *Messiah* oratorio – was coming to pass, 'For cutting through a sand hill this steam digging machine must make a great saving of labor ... Thus the hills are laid low, and the valleys made high!'

This shovel worked on the Western Railroad for some three years and it was during construction of a later extension of the line to Chester that, in 1839 at Westfield, William Otis died at the age of twenty-six.

In 1841 one of the shovels was at work in Troy, New York, and later that year and during 1842 both Otis shovels were operating at Brooklyn, New York, excavating and filling swampland to make the City Park. Justification for the use of the steam shovel was given by members of the American Institute after visiting this site, 'The masses of unruly men collected on our public works will be dispersed by its use, and compelled to till the land, thereby making them good and quiet citizens.' It is likely that one or both machines were used later in the 1840s on the Atlantic Docks project, also in Brooklyn, as the shovels' owner by then, Daniel Carmichael, was a contractor. Subsequent records of the use of the two Philadelphia-built shovels are vague: it appears the last one was broken up during the early 1900s, following a record of one being overhauled in about 1893 at Omaha, Nebraska.

Excavators entered world trade right from their inception. By early in 1842 one was at work in England near Brentwood, Essex, on the Eastern Counties Railway. John Duncan, who had been granted an English patent presumably under some agreement, was responsible for its introduction and John Braithwaite was entrusted with the construction and licensing of further machines for the UK and Europe. There is, though, no evidence of Otis machines being made outside North America. In 1845 one improved by J. B. Hyde was mounted on a scow and used as a dredge in England, and in the following year an Otis shovel was employed at Hull, Humberside, on excavating the Victoria Dock, but these reports may well refer to the original imported unit. Representatives of Russian authorities saw the Otis shovel at work on the Western Railroad in 1839: the result was that the railway company's then consulting engineer, Major George G. Whistler (father of the famous portrait painter) was appointed consultant for the construction of the St Petersburg (Leningrad)–Moscow Railway. The first of four Otis shovels was shipped to Russia, via France, in 1842, with the others following in 1843.

In 1844 Oliver S. Chapman married William Otis' widow, Elizabeth. He had also been a contractor at Canton when William Otis made his prototype excavator, and had been associated with Carmichael, Fairbanks & Otis on the Western Railroad project near Springfield. At the time of his marriage Oliver Chapman was again working with Daniel Carmichael, this time in Vermont, with one or both of the Philadelphia-built machines. Elizabeth Chapman had, of course, control of her late husband's patents – on which she later obtained an extension, from 1853 to 1860. Oliver Chapman secured a large contract in the mid-1850s which prompted him to re-start production of Otis shovels in a modified form, and since that time excavators have been constantly in production. These machines were made for him, probably from the late 1850s, by the Globe Iron Works of South Boston, Massachusetts, owned by John Souther; from 1864 the firm was known as John Souther & Company. Oliver Chapman took out his own patent on improvements for the Otis shovel in 1867, including a chain replacing the shaft driving the crowd mechanism.

Around 1870 an 'Otis-Chapman' shovel was imported into the UK and during the same period several barge mounted versions were used as dredges for further construction work on the Welland Canal; a shovel that also worked on this project later was moved to the Canadian Pacific coast. By 1880, when Ferdinand de Lesseps ordered an Otis-Chapman shovel for his abortive Panama Canal project, Souther had already manufactured over 500 machines. Basically Otis machines, referred to as Otis, Otis-Chapman or Chapman-Souther made under licence by Souther, dominated the excavating field well into the 1880s and were produced until about 1912. Oliver Chapman died in 1877 and is buried in the same plot as William Otis at Canton.

With the death of the excavator's inventor in 1839 and the transfer shortly after of its maker's business to Russia to construct locomotives and rolling stock in St Petersburg, the incipient excavator manufacturing industry had lapsed into dormancy for over a decade. Perhaps not

An Osgood No.1 shovel with distinctive lattice boom working on the Chicago Drainage Canal in the 1880s.

surprisingly, Daniel Carmichael as owner of the only two steam shovels in North America wished to keep the advantage he had over competitor contractors. Nevertheless, the seeds had been sown for the inevitable start of design and manufacture by others.

In 1842 Jason C. Osgood, a contractor of Troy, New York, had met Daniel Carmichael and probably seen an Otis shovel for the first time: the two men then designed and patented in 1846 a two horse (literally) power dredge incorporating Otis shovel principles, and during that year made an agreement for the Starbuck Brothers' Iron Works also of Troy to manufacture these – a company which was eventually to become an excavator manufacturer in its own right. By 1852 Jason Osgood had produced drawings of a 'land excavator' version, again powered by two horses.

Another Osgood, Ralph R., quite likely drawing on knowledge gained when he had· worked with Jason and Daniel Carmichael, took out patents in 1875 and 1877 for an excavator which substituted two legs for the Otis-type mast and replaced its triangular 'crane' with a boom. He formed the Osgood Dredge Company in the latter year; again at Troy, and possibly incorporating Starbuck Brothers, and only three years later the company's products were well enough established for Ferdinand de Lesseps to order the first of eight shovels. By 1882 the firm had become Osgood & MacNaughton and had moved to Albany, New York. This company was exceptional in that it exclusively manufactured dredges and excavators whilst others built such machines as a job-shop operation to supplement a main product line.

Hosea T. Stock changed from being an operator on the early Otis shovels to a manufacturer. The manner of his appearance on the manufacturing scene, however, is unclear. It has been reported that he acquired two Otis shovels in 1851; if so, these must have been manufactured abroad, original machines returned from the UK or Russia, or the two originals disposed of relatively soon by Daniel Carmichael. None of these possibilities appears likely, nor can they be supported by any evidence. Nevertheless, Hosea Stock – who as early as 1844 converted what was possibly the prototype Otis shovel into a dredge and later, in 1861 with William Otis' brother

Isaac, improved an Otis machine by including, for example, iron cladding on the frame – became established as a contractor and excavator builder. His most significant contribution to excavator design was in 1877, when he introduced a steam shovel mounted on a standard-size flat railway wagon: the first 'Wilcox and Stock' railroad shovel. This propelled itself for short distances on standard gauge track, and unlike broad-gauge Otis-type machines could be coupled to a train and easily travel long distances. It could also be converted into a wrecking crane or 'derrick car'. Soon afterwards Hosea Stock introduced independent engines for the three basic digging motions of a shovel. The Toledo Foundry Machine Company, which was founded in 1872 and changed ownership to become the Vulcan Iron Works in 1882, made railroad shovels for Hosea Stock from about 1877 to 1886 and thereafter produced its own 'Giant' series. However, Hosea Stock apparently continued as a contractor and excavator builder until his death in 1891.

By 1880 there were probably only a handful of excavator manufacturers in the USA including John Souther, H. T. Stock, the Starbuck Brothers/Osgood Dredge, the Vulcan Iron Works of Chicago, Illinois and John King & Company of Oswego, New York. Nothing is known of the Vulcan Works of Chicago, but in 1870 the Oswego Boom Machine was patented by C. H. Sage & S. A. Alger and was then made by John King & Company. This was of the luffing-type and was in production for a relatively short time. However, in 1882 S. A. Alger's Friction Excavator was introduced using a vulcanised paper clutch and produced at King's Vulcan Iron Works, Oswego.

In 1880 Dan P. Eells, a mid-West entrepreneur of Cleveland, Ohio, whose bank had been active in promoting the Ohio Central Railroad, was instrumental in establishing the Bucyrus Foundry and Manufacturing Company at Bucyrus, Ohio. This was the location of the new railway's operating headquarters and workshops and the Bucyrus company was created to produce associated railway and mining equipment. Bucyrus stayed there for thirteen years until

Known manufacturing locations in the USA and Canada.

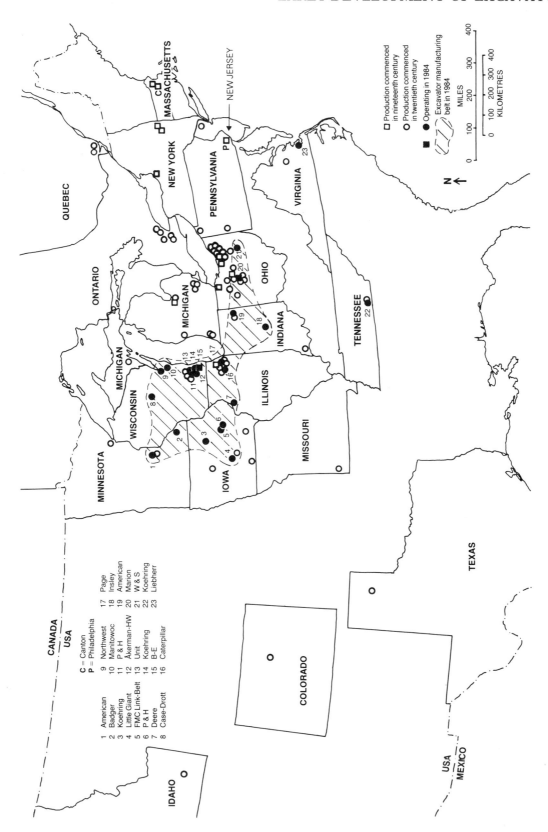

CANADA

USA

C = Canton
P = Philadelphia

1 American
2 Badger
3 Koehring
4 Little Giant
5 FMC Link-Belt
6 P & H
7 Deere
8 Case-Drott

9   Northwest
10 Manitowoc
11 P & H
12 Åkerman-HW
13 Unit
14 Koehring
15 B-E
16 Caterpillar

17 Page
18 Insley
19 American
20 Marion
21 W & S
22 Koehring
23 Liebherr

Production commenced in nineteenth century
Production commenced in twentieth century
Operating in 1984
Excavator manufacturing belt in 1984

moving to its present site in South Milwaukee. In 1882 the first railroad shovel, based on Hosea Stock's design principles and named after its designer, John Thompson, was delivered to the Northern Pacific Railroad. Soon after another went to the Northern Pacific, one to the Ohio Central and one to the Savannah, Florida and Western; an order from Ferdinand de Lesseps in 1883 provided Bucyrus' first export.

In 1884 Henry M. Barnhart, who had been involved the previous year on the construction of the Chicago and Atlantic Railroad, together with his cousin George W. King and Edward Huber, founded the Marion Steam Shovel Company – to the south west of Bucyrus at Marion, Ohio. The first three Barnhart Special railroad shovels were built in Edward Huber's factory in Marion, but the following year the company moved to its own premises nearby. Thus were established within a mere four years of each other and amazingly only 27 km (17 miles) apart, the two companies which were to dominate the international excavator industry in future years: not only this, but their establishment marked the beginning of intense rivalry which has persisted ever since. One Sunday in 1900 the Secretary-Treasurer of Bucyrus climbed over the fence around Marion's premises and made a surreptitious tour of the factory. By the 1920s it appeared the entire Bucyrus sales organisation had a 'Marion fixation' – one employee was taken to task for riding on a train in the company of a Marion salesman! This rivalry has acted as a spur to innovation and improvements in the products of the two companies themselves and been reflected in developments of other manufacturers as well; there has been continual interchange between the two companies of leadership in some aspect of excavator design and production.

In the 1880s and 1890s the number of US excavator manufacturers increased; as did the range of applications, in and around this period. To railway excavation and general public works were added coal stripping in the 1870s, brick yard operations in the 1880s and phosphate and copper mining in the 1890s. The opening up by surface mining of the major Mesabi, Minnesota, iron ore range in the 1890s also created an obvious market for steam shovels. New makes of excavator included the short-lived Clement made

by Industrial Works of Bay City, Michigan (which had cylinders directly operating its crowd and slewing motions) and the Ohio and Victor types. Manufacturing operations tended to move west as demand increased and new products appeared to meet this demand. An indication of the acceptance of steam shovels on major earth-moving tasks was provided in 1894 by the Chicago Drainage Canal project. This reversed the flow of the Chicago River and 54 machines, including 24 Bucyrus, were employed. Ten years later usage on an even larger scale was initiated. Between 1904 and 1914 the Isthmian Canal Commission bought 102 steam shovels, made up of 77 Bucyrus, 24 Marion and one Thew, for construction of the Panama Canal. Out of the 172 million m³ (225 million cu yd) of material moved, 78 million m³ (102 million cu yd) of dry excavation was done within the 14.5 km (9 miles) long Culebra Cut – later Gaillard Cut – on the backbone of the isthmus. This cut alone accounted for about 90 per cent of steam shovel activity in the canal channel up to 1914 and is where two shovels working towards each other met in 1913. When excavation was at its height in 1908 President Theodore Roosevelt boarded an 86 t Bucyrus railroad shovel during an inspection of the 'Big Ditch'. Unforeseen slides in the cut resulted in one shovel being picked up and carried, undamaged, half-way across the site; on another occasion a shovel undertook 103 trips across the toe of a constantly moving slide, keeping pace with the movement of material so that its track did not need to be relocated.

Also in 1904 Bucyrus shipped the first of eighteen shovels to the Rio Tinto copper mine at Huelva, Andalusia, Spain: this, one of the oldest mines in existence, is referred to as 'Tarshish' in the Bible and Roman and Phoenician workers preceded the Bucyrus machines.

As the UK was introduced to the Otis shovel very soon after its invention it was appropriate that the first excavator to be produced outside the USA should be made in England. In 1875 Ruston Proctor & Burton (Ruston Proctor after 1889) built its first steam shovel at Lincoln, Lincolnshire, and delivered it to railway contractors. A year earlier the company had secured patent No. 4480 for an excavator designed by James Dunbar, a Scottish engineer. Although the patent

Ruston & Dunbar excavators were first made in 1875 in Lincoln, Lincolnshire; this one was employed building the Great Central Railway (1894–98) through the centre of Nottingham, Nottinghamshire.

covered a full slew machine, the first Lincoln-built units – like all their counterparts in the USA up to that time – were of limited slew. By the end of 1877 a total of about seventeen 6 kW (8 HP) and 7.4 kW (10 HP) Ruston & Dunbar Steam Navvies had been made. Other than the dipper arm which was made of oak reinforced with mild steel plates, the Ruston & Dunbar was constructed of steel and wrought iron. The Otis-type mast was replaced with a metal tower-frame with its slew ring at the base, and a double cylinder engine of 20 cm (7¾ in) bore was bolted to the front of its vertical boiler. A base frame 3 m (10 ft) wide and mounted on double flange wheels

could be secured by screw jacks at each corner and under the slewing centre. Bucket capacities were up to 1.5 m³ (2 cu yd) and the machine weighed some 30–40 t.

Further UK models of excavator followed the Ruston & Dunbar quickly. In 1877 a patent was taken out by Alexander Chaplin and Company of Motherwell, Strathclyde, for a similar machine; but only two or three were made. In the same year Barclay and Company of Kilmarnock, also in Strathclyde, introduced their version of the steam shovel of which about twelve were produced, and one still worked near Liverpool, Merseyside, in 1927.

In 1884 an excavator that heralded a new era in earthmoving was manufactured by Whitaker & Sons of Horsforth, near Leeds, West Yorkshire, and tested locally at Menston. The most

Whitaker & Sons of Horsforth, near Leeds, West Yorkshire, produced the first full slew excavator in 1884.

significant development in excavator design since the introduction of the Otis shovel occurred when Whitaker created the first 360° slew shovel by fitting a digging attachment to a standard, fully revolving, 3 t Balmforth crane. Thrust for crowding was provided by a hand operated system, but power was added three years later by means of an oscillating ram. At first Whitaker only made the excavating gear but later manufactured the complete machine. In nearby Rodley, the Thomas Smith crane manufacturing company produced its first excavator in 1887, and the same year saw John H. Wilson and Company of Liverpool, Merseyside, attach excavating gear to its 10 t cranes for sale to T. A. Walker, contractor to the Manchester Ship Canal.

In 1880 Ruston & Dunbars were used on construction of the Midland Railway from Melton Mowbray, Leicestershire to Nottingham, Nottinghamshire. Later, from 1884 to 1898 the Great Central Railway was built from north of Nottingham via Leicester, Leicestershire, and Rugby, Warwickshire, to London. Forty-three steam shovels were employed, mainly Ruston & Dunbars but also including some Wilson and Whitaker machines.

The 57-km (35½-mile) Manchester Ship Canal built between 1887 and 1893, from the River Mersey at Eastham, Merseyside, to the new Manchester Docks in Salford, Greater Manchester, probably utilised the largest fleet of steam shovels to operate anywhere to date. Eighty-three British-built shovels, plus seven French and German bucket chain excavators, removed 38 million m³ (50 million cu yd) of dry material. Fifty-eight Ruston & Dunbars with 0.76–1.3 m³ (1–1¾ cu yd) buckets made up the bulk of the fleet; seventeen Whitaker/Smith shovels – which became known as Jubilee shovels because they were first used on the canal in the year of Queen Victoria's jubilee – with Wilson machines made up the rest. In addition, two UK companies later to become excavator manufacturers were represented; Priestman with grabs and Ransomes & Rapier with cranes.

The first fifty or so years of development had seen the excavator – still basically as a shovel – firmly established in two countries as a vital bulk earthmoving tool. Although the fundamental principles had been retained remarkably intact, a range of improvements had already been made to the basic Otis design. Mobility, for example, had

Work stopped on the construction of the Manchester Ship Canal (1887–93) when these ladies braved the mud to pose in front of a Ruston & Dunbar shovel.

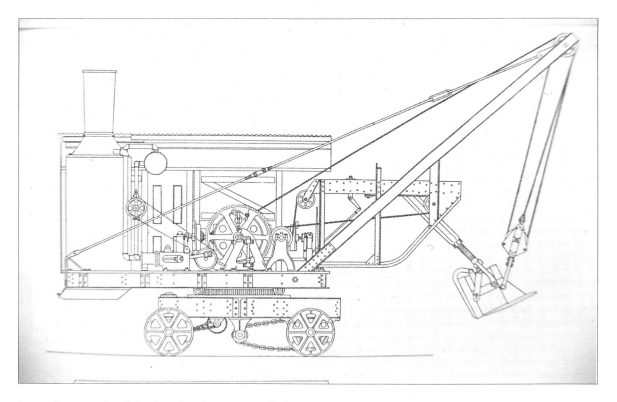

Captain Thew's design for a shovel, first made in 1895, included a level crowd system to prevent damage to wooden planking at Great Lakes' docks.

been increased with the development of the railroad shovel, and versatility enhanced by full slew capability on some models. The next fifty years were to witness both design refinement and rapid progress on a number of fronts. A large number of new manufacturers were to be set up, many in countries other than the USA and UK and sometimes under a licence agreement, and world trade in such machines increased. The inter-related developments of technical changes in digging equipment, methods of mounting and new power sources, together with an ever increasing range of applications, were to feature throughout the period.

The acceptance of the full slew excavator was slow, but when technical problems had been overcome, it had great advantages over earlier machines. It could pick up its own track or mats, required a smaller crew, had a greater digging range and could dig back to its starting point. Although in 1886 an Osgood 0.57 m³ ($\frac{3}{4}$ cu yd) model No. 6 shovel with full slew, designed by John K. Howe, was put on test in the clay pit of a brickyard, it was not until nine years later that this facility became standard on a US-built machine. In 1895 Richard P. Thew, a Great

Lakes ore boat captain, commissioned the Variety Iron Works of Cleveland, Ohio, to build an excavator to his designs. Not only did this have full slew, but it was mounted on broad-rimmed road wheels, and had a novel horizontal crowd mechanism to allow the shovel to clean up stockpiled ore at the docks without damaging the timber planking. Eight Thew-designed machines were built at Cleveland before the Thew Automatic Shovel Company was founded at Lorain, Ohio, in 1899 for their large-scale manufacture. During the first decade of the twentieth century Vulcan, Marion and Ruston Proctor put full slew excavators into production, but it was not until 1912 that the first such machine emerged from Bucyrus.

The Thew Company indirectly gave rise to the growth of another manufacturer whose products really established beyond doubt the role of the small full slew excavator in earthmoving. A. C. Vicary had been an employee of the Thew Company since 1908; in 1913, by which time Thew was the world's largest maker of full slew excavators, he had a chance meeting on a train with Fred McBriar, son of James McBriar who had bought out the original owners of the Ball Engineering Company; manufacturers of stationary steam engines in Erie, Pennsylvania, since 1883. As a result of this meeting A. C. Vicary was employed by Fred McBriar later in 1913 to put the Ball Company in the excavator business. Following a procedure still commonly practised, experience was brought in from other manufacturers: in this case in the form of two additional employees both from Marion but one also with recent experience at Bucyrus. This team gave birth to the first Erie-B 0.57 m³ ($\frac{3}{4}$ cu yd) steam shovel, in 1914, of which 3130 were made out of a total of 4439 excavators produced by the time the Ball Company merged to become Bucyrus-Erie in 1927.

The 'B', and the 0.38 m³ ($\frac{1}{2}$ cu yd) 'A' model, introduced in 1916, were aimed towards general building construction, sewer and road works. The Erie-B, like other Erie excavators, was a full slew machine from the start; it was available with a variety of front-end equipment and mountings, had automatic one-man control, a level crowd facility, and unusually good protection of easily accessible working parts. The Ball company aimed to make its Erie line of excavators the 'Fords' of the industry – a very limited number of standard models, produced in lots and with interchangeable parts. Advanced promotional efforts included the widespread use in advertisements of a mythical character called Bill Muldoon, shovel operator. Mary Pickford, with fellow film star Douglas Fairbanks looking on, took the controls of an Erie shovel in the ground-breaking ceremony for the United Artists Theatre Building in Los Angeles, California.

It was possible for a shovel to dig trenches: on a sewer scheme in Chicago, Illinois, in 1909, a specially mounted Bucyrus 70C shovel with a 16.5 m (54 ft) dipper arm dug down to 7.9 m (26 ft). Nevertheless, the shovel was not ideally suited to digging below ground level. Although in 1880 Osgood patented a backacting (backhoe) type of machine there is no evidence of any being made. At the end of the 1890s, however, both Marion and Vulcan did produce a few primitive backacters – the Vulcan one being patented by James Kewley in 1896. In the early 1900s in the UK Smith sponsored the invention of a Manchester engineer named Jubb who had patented a design for a backacter excavator. Several Jubb Trenchers based on 3 t Smith steam cranes were produced; one worked in the Derwent Valley, Derbyshire, around 1907. Between 1906 and 1909 Whitaker also produced one or more backacters, including a 0.38 m³ ($\frac{1}{2}$ cu yd) machine used to excavate chalk at a Kent cement works. From these early devices backacter equipment developed as a piece of standard equipment for later universal excavators.

The Osgood patent of 1880 also covered the first dragline design, but as detailed on p. 55 it was 1904 before the first practical machine was devised and put into use by John W. Page. The association of John Page with John Monighan followed and the walking dragline was soon created by Oscar Martinson. Bucyrus (just joined by Vulcan) entered the dragline field in 1910, when it began to manufacture the Heyworth-Newman machine devised by James O. Heyworth in 1908 and originally marketed through H. Channon Company of Chicago. The first dragline built in the UK was a Ruston & Hornsby which appeared in 1918 and was used alongside the Manchester Ship Canal by Harry Fairclough, contractors. US-built draglines, though, had been imported into the UK since 1914; these included a Bucyrus Class 14 in 1916 which went to iron ore mines on the small island of Raasay, off Skye, in the Western Isles of Scotland. The rapid development of the dragline as an important form of front-end equipment is illustrated by the use of a fleet of forty-six steam, diesel and diesel electric Ruston & Hornsby and Bucyrus draglines in 1924–32 on the Lloyd Barrage and Canals Construction Scheme in the Sind area of India (now Pakistan) between Sukkur and Hyderabad. To create the main 1402 m (4600 ft) dam on the River Indus behind which a 241 × 72 km (150 × 45 miles) lake was formed, and dig over 9,660 km (6,000 miles) of irrigation canals, machines weighing 20–300 t each were used with bucket capacities ranging from 0.67–7.6 m³ ($\frac{7}{8}$–10 cu yd): these were the first draglines to work in India.

The backacter and dragline were also joined in the early years of this century by another form of digging equipment for excavators, to supplement

This picture of Keystone skimmers at Ocean Parkway, Brooklyn, New York, used to hang in the factory at Beaver Falls, Pennsylvania.

the original shovel – the skimmer. In 1907 Leroy P. Clutter of Washington, Pennsylvania, conceived the idea of fitting a shovel onto a traction engine to produce a light excavator. He found a company to develop the idea but did not meet with initial success. However, in 1912 he approached the Keystone Driller Company of Beaver Falls, Pennsylvania, with designs for which he had made a patent application in 1911. Leroy Clutter's 'Dirt digging and loading device' was the first skimmer; an excavator whose bucket slides along a lowered boom allowing it to skim off a horizontal layer of surface material, particularly useful for roadworks and coal loading. The first machine was shipped to Pittsburg in 1913 but was returned and dismantled. Later that year a second unit was delivered to S. B. Markley in Rochester, Pennsylvania, and put to work on a street grading job. By the end of 1919 725 machines had been built (also convertible to shovel and backacter) and Keystone excavators were made not only as wheel traction engine types but also later as half-tracks and ultimately conventional full-track models. Both steam and petrol engines were used on early versions. Leroy Clutter also devised a telescoping boom skim-

mer, convertible to backacter, which was first produced by the Star Drilling Machine Company of Akron, Ohio, in 1925. In about 1915 Bucyrus produced the 27-B skimmer, designed by William M. Bager, later to become the company's chief engineer, for coal loading in Kansas.

The application of crawlers to excavators, after practical development for tractors, was a logical step: wheeled machines had clear limitations. Those mounted on wheels running on rails had very restricted mobility and as early as the 1880s the 0.38 m³ ($\frac{1}{2}$ cu yd) Vulcan 'Little Giant' shovels had pioneered the use of broad rimmed traction wheels and were self-propelled: this form of mounting was a distinct improvement for any machine undertaking general construction in urban areas. The increased stability and lower ground pressure provided by crawlers, without loss of mobility, were, though, factors that could not be ignored by excavator manufacturers. In 1905 Leach Brothers, contractors of Maumee, Illinois, approached Bucyrus with designs of a crawler mounted excavator, of which they had already built a prototype. Bucyrus made a few of the Leach type with crawlers incorporating planks and steel links forming an endless chain, then in 1911 Bucyrus produced a Class 14 dragline, weighing 62 t, and mounted on four crawlers based on the Leach plus Holt and Best

The Bucyrus 120-B was the first hard-rock quarry excavator on two crawlers and resulted in the demise of railroad shovels in the 1920s.

(who had developed crawlers for tractors in the US) principles. This was the first application of such a mounting to heavy machinery and the initial machine went to work on a drainage scheme near Ormito, Texas. By the 1920s crawler mounting was becoming standard practice with most manufacturers, although Erie did not actively promote this method of mounting until 1923 and Thew even later. In that decade sets of crawlers to convert railroad shovels were marketed by Marion.and Bucyrus: Marion supplied eighteen sets to the Utah Copper Company at Bingham Canyon, Utah.

The final major technical advance in the second half of the first hundred years of excavator development was the introduction of electric motors and petrol or diesel engines and the gradual demise of steam as a source of power. In 1889 an Osgood railroad shovel was powered by two electric motors and in 1899 a 1.1 m³ (1½ cu yd) Vulcan railroad shovel used three electric

motors. Thereafter Vulcan led the way in applying electric power to both limited slew and fully revolving excavators in the early years of the twentieth century. Other manufacturers also began applying electric power for the first time during this period: Ruston Proctor in 1902, Thew in 1903 and Marion in 1908. Electricity gradually became the standard source of power for larger machines such as stripping shovels and walking draglines detailed in later chapters. A Monighan dragline delivered in 1910 was probably the first excavator to be fitted with an internal combustion engine; in 1912 Bucyrus shipped a petrol-engined dragline and in 1921 that company was one of the first to produce a diesel powered shovel. Despite the gradual coming into use of internal combustion engines, steam power held its own for many years.

New manufacturers, or existing manufacturers adding a new product, resulted in many new makes of excavator in the early years of this century. Typical of these was the Atlantic railroad shovel with cable hoist made by a branch of the American Locomotive Company in Richmond, Virginia, and Paterson, New Jersey, as

well as by the Locomotive and Machine Company in Montreal, Quebec. It was designed by A. W. Robinson who left Bucyrus in 1900 and, from 1904 up to the time the Atlantic Equipment Company was taken over by Bucyrus in 1911, over 150 machines had been produced – including three delivered to the UK iron ore fields between 1905 and 1911. Other US makes appearing at this time, but now no longer in existence, included Fairbanks, Browning, Orton, and McMyler Interstate. The German industry was established in 1904; early manufacturers were O & K and Weserhütte, both appearing on the scene in 1908, and Gottwald of Düsseldorf, with its rail mounted VE model grab in 1910, made the

After forty-five years under water this 1909 Ruston Proctor steam crane navvy was restored in 1977–80 and is now preserved at the Museum of Lincolnshire Life in Lincoln, Lincolnshire.

One of a number of railroad shovels built at the beginning of this century by Atlas, now Atlas-Copco, at Stockholm for the Swedish railways.

first step to becoming an excavator manufacturer. In France Pinguely, established in Lyon in 1855, was producing excavators early this century. UK manufacturers emerging included A. R. Grossmith who, in 1908, designed and built a shovel with rope crowd, and a turbine engine next to the bucket to give it variable pitch; Henry Berry of Leeds and Taylor-Hubbard of Leicester built excavators to the designs of Douglas Whitaker, nephew of the original Whitaker Brothers, who had established an office in Leicester – in 1911 Berry built an unusual 'Flip Flop' shovel conceived by Douglas Whitaker on which the dipper arm, fulcrumed at the racking shaft, had a rope attached to its top which was pulled down to create the digging motion; Ransomes & Rapier produced its first excavator in 1914; Rubery Owen of Darlaston, West Midlands, made a single 0.38 m³ ($\frac{1}{2}$ cu yd) 'Handy Navvy' designed by A. R. Grossmith in 1916: and Cowans Sheldon of Carlisle, Cumbria, also made a steam shovel in 1917. Atlas (now Atlas-Copco) in the early years of this century added a few railroad shovels to the range of locomotives built at Stockholm and thus established the excavator industry in Sweden: some units went to the Swedish railways and at least one to the Kiruna iron ore mine in that country's northern Lapland region.

A licensing agreement led to the excavator manufacturing industry being established in Russia when a licence was given in 1900 for Bucyrus-designed shovels to be made at the Poutilov works in St Petersburg. By 1910 five large steam shovels had been constructed: perhaps one of these was the 59 t Bucyrus railroad shovel which worked on the Bulogoe and Polski Railway in 1905. Another licence from 1904 to 1909 gave rise to Canadian-Bucyrus machines being made in Toronto, Ontario. Exports from the USA indicate use of excavators in increasing numbers of countries. For example, in 1912 a Browning shovel went to Mexico; between 1914 and 1916 Bucyrus excavators were delivered to Sweden, the Congo and Chile, and in 1920 to Japan and Manchuria.

The basic pattern of excavator design, manufacture and application was thus well established and being consolidated by the 1920s and 1930s. In particular the internal combustion engine,

crawler mounting and a full slew facility were becoming standard features. Noticeable was the growth of small 'universal' machines capable of taking shovel, backacter, dragline or skimmer equipment as well as having other uses such as those of grab or crane. The most striking feature of the era was the growth of new manufacturers world-wide, many of which are still in existence, and the extensive range of models being created by these new, and by existing makers. In the USA Link-Belt, Northwest, Insley, Koehring and Manitowoc are amongst present-day manufacturers who produced their first excavators in the 1920s: Austin, Massillon and Hanson are examples of makes introduced then but no longer in existence. In the UK Priestman and Smith excavators appeared and Ransomes & Rapier began regular production; Newton Chambers (NCH) and Allen introduced machines made under licence, and Arrol was a manufacturer for a short time. Demag (Germany), Åkerman (Sweden), Kobe and Hitachi (Japan), and Dominion (Canada) excavators appeared, and the USSR established its own excavator industry.

Railroad shovel production ceased; B-E made its last one in 1929. Events such as the giving in 1909 of a $5 panama hat to the driver who filled his train quickest from a Bucyrus 70C were a thing of the past. Most railroad shovels still in use had been converted to electric power. The rope crowd was introduced onto Bucyrus machines in 1922: W. J. Bowtell and R. S. Lewis of Ransomes & Rapier had patented a system which had earlier been sold to Bucyrus. Traction wheels gained a surprising popularity in the 1920s and they were fitted to large railroad shovels over 100 t for quarry and mining operations. At the Kiruna mine a 3 m³ (4 cu yd) Menck & Hambrock full swing excavator of no less than 173 t was mounted on wheels and put to work in 1925. This iron ore mine was a significant user of excavators: its seventeen machines in the 1920s were all electric powered – Sweden has abundant hydro-electric power but lacks coal and oil, and in 1927 a 244 t crawler mounted shovel of 3 m³ (4 cu yd) capacity, was built to the design of the mine's engineers by the Morgårdshammar company from the town of that name in Sweden.

Excavator models of particular interest intro-

Over 7,000 10-RBs were made between 1934 and 1969.

duced in the 1920s and 1930s included the 3 m³ (4 cu yd) 120-B, which appeared in 1925 – the first electric shovel mounted on a single pair of crawlers designed for speed, mobility and hard digging. Even the inventor Thomas A. Edison wrote from Orange, New Jersey, in the year of its introduction for information on the new machine! Although universal excavators had been manufactured for some years previously the 0.76 m³ (1 cu yd) 30-B, dating from 1920, pioneered a trend in such machines, as crawler mounting was standard, and within two years of its introduction steam, petrol or electric powered versions were available. One dug the basement for the Conrad Hilton hotel in Chicago, Illinois. Erie GA2 ('Gas and Air') and DA2 ('Diesel and Air') models were unique excavators brought out in the 1920s and operated by compressed air produced by petrol or diesel engines. A DA2 worked in 1931 on Ford's Tapajoz rubber plantation in Brazil. In 1932 P & H brought out the all-welded 0.29 m³ (⅜ cu yd) 100 model and in the same year Priestman produced the popular 6.6 t Cub. Ransomes & Rapier, which applied fluid couplings to small excavators in the late 1930s,

introduced the even smaller 5.6 t 4½0 in 1937.

In 1930 the merger of two of the oldest excavator manufacturers took place; Ruston-Bucyrus was formed from Ruston & Hornsby and Bucyrus-Erie: B-E being encouraged by Marion's recent licence with Ransomes & Rapier and a desire to expand into the European market. At the time of R-B's creation Ruston had a range of eleven models from the 0.25 m³ (⅓ cu yd) No 3 to the 6.1 m³ (8 cu yd) No 300, which had been introduced in 1923. Only the Ruston No 4 of which 936 were made between 1924 and 1933, was retained: the rest of the range was replaced by 1932 with eight B-E models up to 1.9 m³ (2½ cu yd). A large number of R & H machines were sold to the USSR in 1930–31. One of the early units of the first R–B model to be made, the 0.57 m³ (¾ cu yd) 1030, was used on excavating the foundations for Lewis' department store in Leeds, West Yorkshire.

The 0.29 m³ (⅜ cu yd) 10-B and 10-RB first produced in 1934 exemplified a trend in the 1930s of reducing machine weight for a given bucket capacity and this model was in many aspects a fore-runner of the vast numbers of mass-produced, small, universal excavators which were to be made in the coming years. 3,500 Insley

Built in 1935 by Ruston-Bucyrus this 52-B worked at a cement works in Oxfordshire and is now preserved at Leicestershire Museum of Technology, Leicester, Leicestershire.

Ks and 10,000 Thew-Lorain TL–20s alone were made in 1944–49, both of 0.38 m³ ($\frac{1}{2}$ cu yd) capacity. The 10-B/RB's manufacture was urged on B–E by R–B and, via the 6B, 1020 and 16-B it emerged as B-E's largest selling model from 1935–39. Not only that, but it gave rise to a range of larger and successful models including ultimately the 22B and 22-RB. The 10-B was in production until 1959, and the 10-RB until 1969 – by which time 7,625 10-RBs had been made at Lincoln.

Also in 1930 as well as the formation of R-B, two firms were established, Bruneri in Italy and Poclain in France which, although not excavator manufacturers then, were to have a major impact on the industry in later years. Two other companies made significant decisions in the 1930s: both Ransomes & Rapier and Marion entered the walking dragline market in 1939, thereby doubling the number of manufacturers of this special type of excavator.

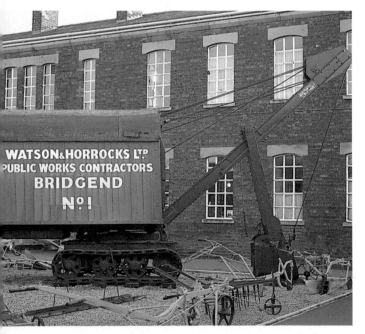

The Ruston No. 4 was a popular machine in the late 1920s and early 1930s; this was the first made by Ruston-Bucyrus in 1930.

CHAPTER 3
Construction~size Cable Excavators

Since the end of World War 2 there have been fundamental changes in the section of the excavator industry covering cable machines up to approximately 7.6 m³ (10 cu yd) bucket capacity. With the birth and growth of hydraulic excavators during these four decades a dramatic decline has taken place in the number of manufacturers providing cable machines and in the overall number of models available. Now only around twenty manufacturers and their licensees or subsidiaries operate in some twenty countries.

The small universal cable excavator has virtually disappeared, unable to compete with backhoe loaders and hydraulic excavators: its general demise was preceded by the earlier loss of one of its popular attachments, the skimmer. Attempts were made by manufacturers, such as Åkerman and Northwest, to combine the best features of both types of machines by cable backhoes having hydraulic rams added to give wrist-action movement to the bucket. This adaptation, though, did not halt the loss of construction-size cable machines. Even the larger sizes of excavators considered here, often used for quarrying and surface mining, have faced increased competition from large hydraulic machines.

Nevertheless, despite the almost complete removal of the small cable shovel and backhoe from the excavating field, and a reduced output and range of larger machines, the dragline – which by its nature must be a cable machine – continues to be made in significant numbers. One hydraulic excavator manufacturer, Liebherr, has actually entered the cable machine field for the first time with a range of draglines as recently as 1981. Even draglines, though, have had to accommodate the development of hydraulic technology: many now use hydraulic rather than mechanical means of power transmission and drive – such as the Smith Eurocrane, Rapier HC and Priestman Lion ranges. There is also limited competition

from what are essentially hydraulic excavators which can be converted to dragline use with the addition of lattice booms and drums, such as the Nobas-Baukema UB1252, and from novel machines such as the Priestman VC15 and Hitachi MA400U.

In the late 1940s and 1950s when the hydraulic excavator was still in its infancy and its likely impact was uncertain; not only did existing cable excavator manufacturers continue to develop models but new makes of cable machines appeared. Landsverk, of Landskrona in Sweden,

A Marion Babcock 43MB working alongside the River Irwell, Salford, Greater Manchester, in 1977.

Krupp-Ardelt excavators were made at Wilhelmshaven, West Germany, from 1955 to 1960.

started production with the LA5 in 1946 and continued to make cable excavators until 1971. The machines had become Kockum-Landsverk in 1965 and hydraulic excavators were also produced from 1966. Also produced in the 1940s and 50s were Swedish Åsbrink and Eksjöverken machines. In the UK, Allen of Oxford, Oxfordshire, developed its own excavators based on the experience of producing Michigan models under licence from 1937 and assembling other US makes – Bay City and Lima – during the war years. These, and other UK machines developed in this period such as Blaw Knox and Bambino (a short-lived product made in Hull, Humberside, around 1959), no longer are made.

The first Brøyt excavator, the 4K of 1952, was cable operated. In West Germany Eder produced its first dragline in 1953; two years later Fuchs entered the cable excavator field and production of Krupp-Ardelt 25RU universal excavators commenced at Wilhelmshaven; Mengele's initial machine appeared in 1957 and Sennebogen started excavator production with a dragline in the early 1960s. Whilst some new excavator manufacturers eventually left this area of activity entirely, such as Allen, Krupp-Ardelt in 1970 and Kockum-Landsverk in 1973, others such as Brøyt, Mengele and Eder replaced cable models with hydraulic ones, and yet others such as Fuchs

and Sennebogen extended their product range by adding hydraulic to cable machines.

The era, however, was characterised by the disappearance of many long established makes of cable excavator: early casualties in the USA were Orton in 1945, Keystone in 1949 and more recently Michigan in 1959 and Lima in 1981 which finished making excavators altogether. Present manufacturers in that country which ceased production of cable machines include Unit which had stopped by 1970, and Koehring which, although still producing draglines, made the last Schield-Bantam designed cable excavator in 1973.

The loss of Lima excavators is particularly noteworthy as the most recent model, the 2400B 4.6–6.1 m³ (6–8 cu yd) shovel or dragline, first produced in the 1950s and becoming a B version in the early 1960s, had earned itself an enviable reputation. Lima Locomotive works was the original manufacturer; this had become the Baldwin-Lima-Hamilton Corporation in 1950 when a competitor locomotive manufacturer was incorporated. The following year another producer of excavators, the Austin-Western Company of Aurora, Illinois, was taken over by B-L-H. From 1971 until production of Lima excavators ceased ten years later the company was owned by Clark Equipment. In 1979 an upgraded 2400B prototype shovel, the 9.7 m³ (12¾ cu yd) LS version, had been put on test in Alabama.

Other US makes ceased to exist through amalgamation. Marion took over Osgood-General in 1954 and later discontinued production of construction-size excavators altogether; Thew, which had itself purchased Byers in 1953, in turn became part of the Koehring organisation in 1964 and now Thew's 'Lorain' brand of cable excavators are no longer in production. In 1961 Unit incorporated Bay City and yet another name disappeared.

Outside North America the pattern was repeated. In West Germany manufacture of Wilhag, Gross, Gottwald and Menck cable excavators finished – Menck after incorporation into Koehring. O & K and Demag phased out cable machines in the 1960s as production of hydraulic replacements expanded; although Demag produced a single dragline, the 410-LC/

In the late 1950s O & K built the L351 cable excavator.   *Demag's largest cable excavator, the electric U35 (1941–44) with the diesel K22 (1938–50).*

Excavating material at Atherton, Greater Manchester, in 1965 for a M62 Motorway link road.

LCB until recently. Similarly in Sweden, the last Åkerman cable excavators, 751 and 752 models, were made in 1969; with the exception of the M14 pile driver made until 1974.

Cable excavators no longer produced in France include those of Pinguely, an excavator manufacturer dating back to the industry's steam era. Pinguely had been part of the Creusot-Loire Group since 1970 when it was joined in 1975 by another French excavator manufacturer, Haulotte, founded in 1924. Creusot-Loire's last excavators made under the Pinguely name up to the late 1970s at St Chamond, near St Etienne, Loire, was a range of GT series draglines. Other

French makes that have gone are Bondy (which also marketed Soviet excavators) and Nordest, machines made within the Richier organisation which were superseded by Oleomat-Ford-Richier hydraulic excavators. In Italy, for example, Comet cable excavators are no longer manufactured, and Fiorentini deleted its last cable machine from its product line, the FB780CL/CLS, around 1980. The Dutch Hovers-HCT made in Tilburg, and the Belgian Boom from the town of that name, are both names of the past.

The final general feature of this section of the excavator industry during the time considered here was the establishment of licensed manufacturing of cable machines. Sometimes this was for a period now completed, sometimes continuing to the present day under the original licence, or its extension, or continuing in practice with machines derived from those made when a licence was in force. Through this procedure construction-size machines, no longer produced by the designing and licensing company in its home country, are still produced – often in under-developed and developing countries lacking adequate local technology. Marion-designed 93-M, 101-M and 111-Ms are still produced in India, and P & H 315, 320, 955A and 955ALC models are made in India and Brazil.

Previous licensing agreements for the manufacture of cable excavators in this class include those given by Koehring to Armstrong-Holland in Australia (1945–81) and to Kynos in Spain (1952–74): by Åkerman to Rimas in Denmark (*c.* 1950–63) and to Varkaus in Finland (*c.* 1950–67); P & H to Rheinstahl in West Germany in the 1950s; Marion to Babcock in the UK (1959–mid-70s) and to Creusot-Loire in France (1960–75); and from Priestman and Menck to Warynski in Poland in the 1960s.

Some early Soviet cable machines, such as the E-153, SE-3 and EKG-4, from around the 1950s, have gone out of production, but many existing models are long-established machines which have been up-dated over the years. The E-302 was in existence certainly in the early 1960s: in 1974 for example thirty wheeled E-302Bs and five crawler E-304Bs all of 0.4 m³ (½ cu yd) capacity were shipped to Ateliers de Bondy in France for marketing. The E-304V, a 0.4 m³ (½ cu yd) dragline version was still in production in 1978.

Dating back to at least 1960 is the E-651. The E-652B, as a 0.8 m³ (1 cu yd) dragline, has been made in large numbers. Between 1968 and 1972 over 1000 E-652 and E-652Bs were exported to Iraq for drainage and irrigation schemes and a further thirty went in 1974: in 1978–81 twenty-four were delivered to the Ministry of Agriculture in Greece. The E-10011A is a universal 1 m³ (1¼ cu yd) machine weighing up to 35 t, and the EO-5112A is also of the same capacity and weight but only available as an electric powered shovel which is adapted for tunnelling and underground mining. This is especially used in exploiting copper ore, gypsum, salt and potash deposits.

A Dutch-made Hovers at work in a typical landscape of its native country.

Chinese excavators include the WK-2.

With weights up to 41 t the universal E-1252B existed as an E-1252 in 1962 and was in its 'B' version by the early 1970s; there is also an E-1251B. The E-1252B is available as a 0.5–1.5 m³ ($\frac{5}{8}$–2 cu yd) dragline, 1.4 m³ ($1\frac{7}{8}$ cu yd) backhoe or 1.2 m³ ($1\frac{5}{8}$ cu yd) shovel. Units of this model have worked in the Sudan and Pakistan and in 1978–79 fifteen were added to the fleet of E-652Bs in Iraq.

Electric drive is used for the E-2503 weighing up to 94 t and in production during the 1960s and 1970s. Normal bucket capacities range up to 3.2 m³ ($4\frac{1}{8}$ cu yd) as shovel or 3 m³ ($3\frac{7}{8}$ cu yd) as dragline with booms up to 25 m (82 ft). An E-2505 version has worked in temperatures down to −60°C. In 1966–73 E-2503s went to Bulgaria, Romania, Vietnam and Yugoslavia and more recently in 1980 to Syria. An E-2503 shovel made in 1966 was still working in a limestone quarry at Effingen, East Germany, 15 years later: its output had averaged over 350,000 t annually.

Remaining Soviet excavators are essentially quarry and mining shovels, although relatively small; they range upwards from the 132 t EKG-3.2 of 3.2 m³ ($4\frac{1}{8}$ cu yd) capacity. The EKG-4, dating from at least 1960, has been replaced by the EKG-4.6 of 4.6 m³ (6 cu yd) and later 'B' version weighing 199 t. A total of 124 EKG-4.6Bs were delivered in 1966–73 to Bulgaria, Romania, East Germany, Vietnam, Yugoslavia, Turkey and India; by 1979 Coal India operated 120 of these excavators. Since 1978 further machines have gone to Romania and East Germany and in 1978–81 nine were shipped to work at Dvur Kralove, Prahovice, Czechoslovakia. EKG-4.6Bs are made under licence in India and China.

In the mid-1970s the EKG-5 was developed as an addition to the EKG-4.6B; this became the EKG-5A in about 1980. The current model, weighing some 200 t and with bucket capacities of 5.6–6.3 m³ ($7\frac{3}{8}$–$8\frac{1}{4}$ cu yd), has lower works propel and a semi-automatic digging action. An annual output of over 2 million m³ (2.6 million cu yd) of material is possible. As well as in the USSR, EKG-5As are at work in East Germany, Yugoslavia, Romania, Cuba, India and Vietnam.

China's cable excavators in this size class include two models of shovel with capacities up to 2.5 m³ (3¼ cu yd) made at Hangzhou. The 1 m³ (1¼ cu yd) W1001, WK-2 and 4 m³ (5¼ cu yd) WK-4 shovels, and W200A/WD200A shovel or dragline, were recently in production.

Hitachi is the only Japanese manufacturer not producing cable excavators under licence, or derived from other makers' designs following licensing agreements. Hitachi made a 'U' series of mechanically driven universal excavators from 1949 until recently: one of the larger models, the U161, was a 136 kW (183 HP) machine with capacities up to 1.6 m³ (2⅛ cu yd). The current KH series of hydraulically driven draglines, the 75, 100-2, 125-2, 150-2 and 180–2 have capacities up to 1.4 m³ (1¾ cu yd) and weights up to 46 t. Nissha draglines with capacities up to 1 m³ (1¼ cu yd) used to be manufactured in Japan but may well have been discontinued.

In Czechoslovakia universal cable excavators such as the diesel engined D063 or its D063E electric version, with bucket capacities up to 1 m³ (1¼ cu yd) were previously produced. It appears only the electric Unex E303, developed probably from the Skoda E302, is now in production. Shovel bucket sizes go up to 5 m³ (6½ cu yd) and as a dragline it can take a 1.7 m³ (2¼ cu yd) bucket with booms up to 25 m (82 ft).

Two factories in East Germany produce cable excavators. Zemag at Zeitz manufactures the Takraf UB1412. This has capacities up to 2.5 m³ (3¼ cu yd) as a shovel, 2.4 m³ (3⅛ cu yd) as a backhoe or 2.3 m³ (3 cu yd) as a dragline: it is powered by a 150 kW (201 HP) twelve cylinder diesel engine. Nobas at Nordhausen produced universal excavators such as the UB80 with 1–1.3 m³ (1¼–1¾ cu yd) capacities as a shovel: 150 units of this model went to Brazil in 1964. Under the Baukema name the factory currently makes the 115 kW (155 HP) UB1252 hydraulically driven dragline which is a variant of the UB1232 hydraulic backhoe or shovel.

Five West German companies at present make cable excavators, plus Liebherr which manufactures its draglines in Austria. The Bavaria range of hydraulically driven draglines comprises three machines, the largest of which is a 14 t machine with a 0.8 m³ (1 cu yd) bucket on a 15 m (49 ft 2 in) boom. Sennebogen market four basic models of hydraulically driven draglines – the 815, 1020, 1225 and 1240, each with M-wheeled, or R-crawler, mountings. Weights go up to 30 t and bucket capacities up to 1.5 m³ (2 cu yd). Five models of Fuchs draglines are currently in production. These are the 107 introduced in 1979, the 110 in 1978, the wheeled 114M in 1977 and crawler 114R in 1978, the wheeled 118M and crawler 118R in 1977, and the largest and most recently introduced machine, the 125R. This dates from 1980, weighs up to 34 t and has bucket capacities up to 1.5 m³ (2 cu yd).

Weserhütte produce two series of crawler excavators: the 'W' series of mechanically driven universal machines was started in 1964 but the W40 and W80 are no longer available. Weserhütte's smallest current 'W' model is the W100 shovel/W120 dragline introduced in 1967; the W160/W180 follows, introduced in 1968 with the W270/W320 being the largest. This was introduced in 1976 with weights up to 101 t and bucket capacities up to 3.8 m³ (5 cu yd). Current models in the 'SW' series of hydraulically driven draglines have been introduced since 1978; the crawler SW80 and wheeled SW80-M have been withdrawn. Weights of the SW140, SW190,

The small Priestman Cub V, made 1956–62, was useful for road works in congested urban areas when equipped as a skimmer.

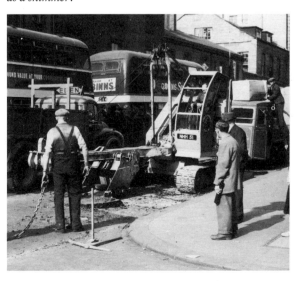

EXCAVATORS

The Smith E1400 dragline has capacities up to 0.5 m³ ($\frac{5}{8}$ cu yd).

SW310 and SW530 range from 40 to 200 t and bucket capacities from 0.8–6.1 m³ (1–8 cu yd).

In 1981 Liebherr's first dragline, the 85 t HS870, appeared, followed in 1982 by the 69 t HS850 and in 1983 by the 51 t HS840. All use Daimler-Benz diesel engines and are hydraulically driven with extensive use being made of electrical and electronic controls. Standard bucket capacity of the HS870 is 3.4 m³ ($4\frac{1}{2}$ cu yd) with booms up to 30 m (98 ft).

Priestman, Rapier-NCK, Smith and Ruston-Bucyrus are the four current manufacturers of cable excavators in the UK; part of the Rapier-

NCK-Rapier 1405B shovel operating at a cement works quarry at Kensworth, near Dunstable, Bedfordshire.

NCK range is based on Koehring-designs and R-B machines are basically the same as those of Bucyrus-Erie. During the last forty or so years Priestman developed a number of versions of several models of excavator such as the 0.29 m³ ($\frac{3}{8}$ cu yd) Cub V introduced in 1957 which became the Cub VI in 1963, the Wolf series of similar capacity and the 0.38 m³ ($\frac{1}{2}$ cu yd) Panther series. The Lion series of mechanically driven machines has now been developed into three models of hydraulically driven draglines – the 40-H, 50-H and 85-H. The Lion 40-H and 50-H have 1.2 m³ ($1\frac{1}{2}$ cu yd) buckets, while the 85-H can take a 3 m³ (4 cu yd) bucket.

Smith's current range of mechanically driven draglines consists of the E1400, with capacities up to 0.5 m³ ($\frac{5}{8}$ cu yd), E2800 and L/W version up to 1.2 m³ ($1\frac{1}{2}$ cu yd), the 25E – introduced in 1980 – up to 1 m³ ($1\frac{1}{4}$ cu yd), and the 4000 L/W up to 2 m³ ($2\frac{1}{2}$ cu yd). The HS35E hydraulically driven dragline with capacities up to 1.2 m³ ($1\frac{1}{2}$ cu yd) was launched in 1976. A shipment during 1984 was particularly interesting; that of a 25E to the remote island of Tristan da Cunha in the South Atlantic. This was to join another Smith machine, an E2600 supplied in 1969, and will discharge barges and also dredge the island's harbour with grab and dragline equipment. An E4000L/W has recently been put to work in a sand and gravel quarry in Essex fitted with a 156 kW (209 HP) Rolls Royce diesel engine.

Original Rapier-designed construction-size cable excavators were phased out in the 1960s and replaced by Koehring-designed machines; however, in recent years British designs have re-appeared in the form of two hydraulically driven draglines, the Orion HC80, introduced in 1984 and derived from the Ajax HC75 launched in 1980, and the Olympus HC170, introduced in 1980 and derived from the HC150 which was launched three years earlier. There are thirteen crawler models in the current Rapier-NCK range; nine of which as excavators are available only as draglines (the Pennine C40, 406D, Andes C41B, Europa C50, Ajax C75, Ajax C75E, Orion HC 80, Eiger C120 and Olympus HC170); three can be equipped as dragline or shovel (305B, 605-2B and 1405D); and the 406 can be used as a dragline, shovel or backhoe. At 26 t the 305B is the smallest model with shovel capacities of 0.57

or 0.67 m³ ($\frac{3}{4}$ or $\frac{7}{8}$ cu yd) and dragline capacities of 0.38–1.3 m³ ($\frac{1}{2}$–$1\frac{3}{4}$ cu yd). In contrast the largest machine, the Olympus HC170 weighs 134 t and employs either a Dorman or a GM Detroit diesel engine of 242 kW (324 HP) driving three variable and one fixed displacement hydraulic pumps. Dragline booms up to 39 m (128 ft) are possible and bucket capacities range from 1.5 to 3.8 m³ ($1\frac{1}{2}$–5 cu yd). An HC170 dragline began operating in 1983 on dam construction in Papua New Guinea. In 1979–80 eighty 305B and forty-six 406D draglines were shipped to Egypt for work in the Nile delta region and upper Egypt; 1405B and 1405C shovels work at a cement works quarry at Kensworth near Dunstable, Bedfordshire, and a 1948 vintage Rapier 410 shovel was still working in Malaysia in 1979.

The last UK manufacturer, Ruston-Bucyrus, and the first of eight US manufacturers, Bucyrus-Erie, can be considered together as R-B excavators have been basically B-E designs since amalgamation in 1930. In 1983 four basic models were offered by both R-B and B-E: R-B produced the 30 Series V standard as shovel or dragline and HD-LC as dragline, B-E made the backhoe and Hi-walker backhoe versions, and both made the HD dragline; R-B manufactured the 38 Series II standard as shovel or dragline and both made the HD version as dragline; R-B manufactured the 61 Series II standard as dragline, B-E made the backhoe version, and both made the standard as shovel and HD version as dragline; and R-B made the 71 Series III standard as dragline, with both making the standard as shovel and HD version as dragline. In addition R-B's line also included the 22-RB ICD shovel and dragline and HD version as dragline – the 22-B has not been made by B-E since 1976; and the 110-RB shovel and dragline – a model which has not been made by B-E since 1975. B-E's line also included the 25-BHD Series III dragline, the partly hydraulically driven 65-D dragline, and the 88-B Series IV as standard shovel or as loading shovel introduced in 1982, and HD version as dragline. The 22-B is also made by FNV in Brazil.

This very wide combined range of machines – reduced in number in 1984 – had shovel bucket capacities ranging from 0.57 to 0.76 m³ ($\frac{3}{4}$–1 cu yd) on the 22-RB to the 88-B loading version

This 38-RB backacter worked in 1963 alongside the Torrens River in South Australia.

and 110-RB both with bucket capacities up to 7.6 m³ (10 cu yd), and machine weights extended from 25 t to 165 t. The 22 dates back originally to 1937, the 30 in basically its present form to 1956, the 38 to 1945, the 61 to 1963, the 65 to 1975, the 71 to 1954, the 88 to 1946 and the 110 to 1950. By the end of 1983 Ruston-Bucyrus alone had made 10185 22-RBs since 1950 (including truck mounted crane versions), 2067 30-RBs since 1956, 1788 38-RBs since 1948, 165 61-RBs since 1967, 220 71-RBs since 1961 and 152 110-RBs since 1955.

The 22-RB has clearly been a highly successful model: for example in the mid-1970s forty-four went to Egypt and thirty-nine to Iraq; with others going to Guyana, Dominica, Trinidad, Tahiti and Puerto Rico around this time. In 1974 three 30-RBs were shipped to Nauru Island and five to Christmas Island, in the Pacific Ocean, for phosphate mining. A 38-RB has even been depicted on a Christmas Island 40 cent postage stamp!

61-RBs went to South Korean and Guatemalan cement works, and worked in Qatar, in the 1970s. 71-RBs during the last twenty years have been employed on works as diverse as dam construction in Argentina and Brazil, road building between France and Spain at the foot of the Pyrenees and sand extraction in South Africa from which gold, silver and iron pyrites are recovered.

In the late 1970s an 88-B together with at least two 71-Bs, three 30-Bs and a 61-B, was operating on the Tennessee-Tombigbee Waterway project in Mississippi and Alabama. Three 88-Bs work at the Minos do Camaque copper mine in Brazil. In the late 1960s three 110-RBs were amongst the fleet of some forty-two Ruston-Bucyrus excavators (including three 150-RBs, four 71-RBs, and 38, 30 and 22-RBs) constructing the M62 Trans-Pennine motorway between Lancashire and Yorkshire. This includes the 49 m (160 ft) deep Deanhead cutting, the 43 m (140 ft) high Rakewood viaduct and the 625 m (2,050 ft) long Scammonden embankment and dam.

Although Marion construction-size excavators are no longer produced in the USA, three models based on Marion designs are still made in India. Both the Hind 93-M (up to 73 t) and 101-M (up to 80 t) were first made in 1961 and are available as shovel, backhoe or dragline with overall capacities up to 2.3 m³ (3 cu yd); the Hind Marion 111-M (up to 120 t) introduced in 1978, can be equipped as a 3.8 m³ (5 cu yd) shovel or 1.3–3.8 m³ (1¾–5 cu yd) dragline: larger coal loading shovel buckets are available. By the end of 1983 a total of 541 units of the three models had been manufactured.

P & H, and P & H-derived, excavators in this class are made in Japan, Brazil, Iran and India as well as in the USA. In the USA in 1983 P & H were offering four crawler dragline models, the 670WLC – a 1.5 m³ (2 cu yd) machine first introduced in 1968, the 5060, 5100 and 5250 Delta introduced in 1983; together with the 325-TC, 430A-TC and 440-TC truck mounted draglines. Crawler versions of two of the latter, the 325 and 440A-S, are made as Kobelco excavators in Japan as well as the crawler 320H and 335A-S. In Brazil (by Villares) and India (by Tata) the crawler 315, 320, 955A and 955A-LC with overall bucket capacities up to 2.7 m³ (3½ cu yd) are made; only as draglines by Villares, but as draglines, backhoes and (with the exception of the 320) as shovels by Tata. Villares also produces the crawler 525 and 535 and truck mounted 425-TC and 650-VTC draglines. Tata also produces the 655B and 1055B as dragline, backhoe and shovel, and the 655B-LC and 1055B-LC as dragline and backhoe.

FMC-Link Belt crawler draglines are made in the USA, Italy, Japan and Mexico: these machines have Speed-o-Matic power hydraulic

systems controlling machine functions. The full current range of models starts with the LS-68 originally introduced in 1954, through LS-78 (and PL backhoe version), LS-98 (and PL backhoe version), LS-98A, LS-108B, LS-108C, LS-118, LS-128DL (and DLC version), LS-318, LS-338 and LS-418A models to the largest, the LS-518 first introduced in 1968. Weights range from 18 to 117 t.

Koehring now only produces four models of crawler dragline: the 305 dating from 1955 has 0.57–0.96 m³ ($\frac{3}{4}$–$1\frac{1}{4}$ cu yd) capacities; the 405, 1953, 0.76–1.2 m³ (1–$1\frac{1}{2}$ cu yd); the 440 Spanner, 1970, 0.48–1 m³ ($\frac{5}{8}$–$1\frac{1}{4}$ cu yd); and the 770 Spanner, 1963, 0.76–1.3 m³ (1–$1\frac{3}{4}$ cu yd). IHI's Koehring design-based range, made in Japan, comprises the K250 and K400A draglines, the 1000 available as dragline, shovel or backhoe and

Draglines are sometimes used to load material into trucks.

the 3.1 m³ (4 cu yd) 1405 backhoe or shovel, with its 1495 dragline version, made to special order.

American draglines in this size range consist of four series of machines: the 400 Series (4120,) 500 Series (597C, 599A and 5299A); 700 Series (797C, 7220, 7250, 7255 and 7260); and 900 Series (998C, 999C, 9260 and 9270). A 4120 worked on a drainage project in Hutchinson, Minnesota; a 597 has loaded dump trucks at Baton Rouge, Louisiana, on a landfill project; a 797 excavated a drainage system at Jacksonville Beach, Florida; and a 9260 strips overburden at an Indiana coal mine.

Although six current Manitowoc crawler machines can be used as excavators – the 3900 and W

Underwater excavation is often undertaken by draglines such as this 61-RB.

version, 4000W, 4100W and 4600 Series 1 and 3 – only the 4600 Series 1 is now normally produced as such, with dragline equipment. Designed primarily as a 3.8 m³ (5 cu yd) excavator the 4500 was Manitowoc's first new model to be introduced after World War 2. The initial 4500 built in 1947 was still at work as a dragline at a coal mine in the Sherman Mountains, Pennsylvania, thirty years later. In 1951 a long-range shovel version was added and in 1958 the Vicon (variable independently controlled) system of power transmission appeared on the 4500. In 1961 the 4500 was upgraded to the 4600 Series I with capacities up to 5.7 m³ ($7\frac{1}{2}$ cu yd) as dragline or 6.1 m³ (8 cu yd) as shovel. Unusually the rope crowd and retract drum of the shovel is mounted on the boom near its base. The current 4600

Series I dragline, normally powered by two Cummins diesel engines totalling 481 kW (646 HP), weighs some 177 t and has an interlock system which eliminates drag and hoist brake riding. In 1984 Manitowoc launched the 3–3.8 m³ ($3\frac{7}{8}$–5 cu yd) 3950D dragline.

Around 1981 at least nine 4600 draglines were working at coal mines in Clearfield County, Pennsylvania. In Florida two 4600s excavated more than 1.1 million m³ (1.4 million cu yd) of material to create recreational lakes for a 40,000 population community being developed near Miami, and in that state one contractor operates ten such machines on road, canal and levée construction projects.

Recently the Northwest range of cable crawler excavators consisted of the 2.3 m³ (3 cu yd) 80-D shovel and the 180-D Series II shovel with bucket capacities up to 7.7 m³ (10 cu yd) as a coal loader; the 3 m³ (4 cu yd) 9570-DA and 5–6.3 m³

(6¼–8¼ cu yd) 190-DA Series II Pullshovels (cable backhoes with hydraulic wrist action on the bucket); and 50-D Jobmaster, 9570, 190-DWT Special Jobmaster and 190-D Series II Heavy Duty draglines with overall bucket capacities up to about 4.2 m³ (5½ cu yd). The 180-D and 190-D were introduced in 1962, the 50-D in 1968 and the 9570 in 1970.

The 80-D was originally introduced in 1933 as the 80 and the 'D' was added in 1937: by 1981

Two Manitowoc 4600s and a 6400 at a coal mine near English Center, Pennsylvania.

some 2600 units had been made, basically unchanged since inception and of which at least 1500 were then still operational. Three 80-D shovels dug granite for the Alcantara Dam in Spain, and in contrast in 1981 an 80-D dragline dredged out a pond in Manhattan's Central Park, New York.

CHAPTER 4
Cable Mining Shovels and Draglines

Large cable mining shovels and draglines comprise two broad groups of excavators. Stripping shovels have not been produced since 1971, superseded by large walking draglines. However, stripping shovels are of special interest because of the mammoth size to which they grew and because many are still in operation on coalfields in the USA. Electric mining shovels, and their large crawler dragline derivatives (machines in excess of approximately 200 t) are, however, very much in evidence as current models and form a group likely to be in production and further developed for the foreseeable future.

In many ways the first true stripping shovel, from which evolved the huge machines of the 1950s and 1960s, was introduced in 1911. Grant Holmes and W. G. Hartshorn conceived of a full swing and self-propelled shovel with long-range digging and dumping capabilities. After initial reluctance when the idea was put to Marion, the company did build such a machine; the first Model 250. This had a 2.7 m³ (3½ cu yd) bucket on a 19.8 m (65 ft) boom with a 12.2 m (40 ft) dipper arm. Straight away this steam shovel exhibited what was to become a distinctive feature of stripping shovels – suspension at four points and a method of levelling the machine for operation. The 250 had hydraulic jacks at each corner to prevent twisting of the lower frame, and it was propelled on four bogies, each with four wheels, which ran on rails. A lattice boom incorporated an inside dipper arm. The first 250

Surface coal mining by stripping shovels was effectively initiated by the Marion Model 250 in 1911; the following year this unit was working with a Model 28 near Sponsler, Indiana.

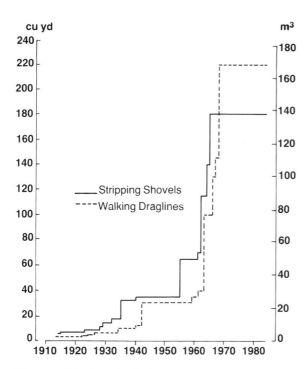

cu yd

m³

Stripping Shovels
Walking Draglines

1910 1920 1930 1940 1950 1960 1970 1980

The growth in size of stripping shovels and walking draglines.

worked clearing 6–9 m (20–30 ft) of overburden at Mission Field near Danville on the Illinois bituminous coalfield.

Bucyrus reacted swiftly. Within a year a competitor to the Marion 250 was designed and built, the Bucyrus 175-B with a 2.7 m³ (3½ cu yd) bucket on a 22.9 m (75 ft) boom available with steam or electric power. A smaller steam model, the 150-B, with a 1.9 m³ (2½ cu yd) bucket on an 18.3 m (60 ft) boom was also produced. Both models were mounted on rail wheels and had a three-point suspension system with screwjacks, although in later years Bucyrus switched to hydraulic jacks. The initial 150-B was bought by the C. F. Markham Coal Company of Fuller, Kansas, in 1912.

Although it was in the USA where stripping shovel technology developed apace, the basic concept had not been ignored elsewhere. In the UK, for example, at the turn of the century the Wilson company of Liverpool, Merseyside, produced a 1.2 m³ (1½ cu yd) stripping shovel with a 21.3 m (70 ft) lattice boom for the mines of Lloyds Ironstone Company near Corby, North-

amptonshire. Similarly, the Whitaker company of Horsforth, near Leeds, West Yorkshire, in 1909–10 made a stripping shovel for use at Scunthorpe, Humberside; again on an iron ore field. It was apparently not, though, until the 1920s that a manufacturer outside the USA adopted the four point suspension system.

The Marion 250 and Bucyrus 175-B and 150-B models initiated a range of new and ever larger stripping shovels – such as the 3.8 m³ (5 cu yd) Marion 271 in 1915 and 4.6 m³ (6 cu yd) Bucyrus 225-B in 1916. Some of these early US-built machines soon crossed the Atlantic to join British-built excavators on the iron ore fields. In 1916 Lloyd's put to work a 193 t 175-B and two years later added a 305 t 225-B. In the early 1920s Ruston & Hornsby No 300 stripping shovels, similar to the 225-B in size, joined these on the UK iron ore fields, and a smaller R & H model, the No 135, soon followed.

1919 was a significant year in excavator design for then both Marion, on a 300-E, and Bucyrus, on a 225-B, applied for the first time the Ward-

'The Mountaineer' – drawn with 'Paul Bunyan' from Michigan! – with its 45.9 m³ (60 cu yd) bucket, erected in 1956, paved the way for the gigantic machines of the 1960s.

The 99 m³ (130 cu yd) B-E 1950-B, introduced in 1965, was the last model of stripping shovel to emerge from Bucyrus-Erie: 1950-Bs were the only B-E machines to have knee-action crowd. 'The Silver Spade' is near Cadiz, Ohio.

A 3850-B stripping shovel would tower above Lincoln Cathedral, Lincolnshire.

Leonard rectified control system to their electric powered machines. Extensively used on electric draglines and shovels ever since, the system gives speed and power curves practically identical to those of a steam engine and better than those of an internal combustion engine. As nearly constant torque results, irrespective of speed which is adjusted to the resistance encountered, it is well suited to excavator use. Induction motor generators (MG sets) driven by a synchronous (constant speed) electric motor supplied with outside AC current, in turn produce power for Ward-Leonard controlled DC electric motors operating hoist, crowd and swing functions. When motors are used as brakes they act as generators, thereby putting power back into the electrical system.

Although recently introduced, by 1927 crawler-mounting was standard on virtually all new stripping shovels. In 1928 9.2 m³ (12 cu yd) machines appeared from Bucyrus and Marion; Bucyrus' model, the 750-B, from inception featured twin hoist ropes and later had a counter-balanced hoist to aid digging. A counter-

The world's largest shovel; 138 m³ (180 cu yd) Marion 6360 at Captain Mine, Illinois.

balanced hoist was also a feature of the 22.9 m³ (30 cu yd) 950-B introduced in 1935, the 26.7 m³ (35 cu yd) 1050-B of 1941 and of Marion's 13.8 m³ (18 cu yd) 5560-E of 1932.

From 1934 to 1942 a total of thirteen Rapier 5360 type stripping shovels, based on the Marion 360 dating from 1923, were set to work in UK iron ore mines. These 5360s weighed over 600 t and had 6.1–8.4 m³ (8–11 cu yd) buckets; later more powerful units with longer booms and dipper arms were designated 5367s. A single 2.7 m³ (3½ cu yd) 5160 of similar design was built in 1938 for a cement works in Oxfordshire. In the early 1960s Rapier did design work on 15.3 m³ (20 cu yd) S1000 and 22.9 m³ (30 cu yd) S1500 stripping shovels, but none were built.

Marion's 'knee-action' crowd device first appeared on a 5561 model in 1940 which was delivered to the Tecumseh Coal Company, Boonville, Indiana. This device, which relieved the boom of weight and stress caused by the dipper arm and bucket, assisted the breakthrough to even larger stripping shovels beginning

with the 45.9 m³ (60 cu yd) Marion 5760 in 1956. The first of five 5760s, 'The Mountaineer', went to a coal mine near Cadiz, Ohio. Greenville, Kentucky, was the destination for B-E's 1650-B 'River Queen' of 1961: this model could take buckets up to 53.6 m³ (70 cu yd). In 1962 the first of B-E's largest model of stripping shovel, the 3850-B, started operating at Peabody Coal's Sinclair mine, Kentucky; this had an 88 m³ (115 cu yd) bucket. The second, and last, 3850-B machine which followed two years later at another Peabody mine, River King near Marissa, Illinois, weighed 8300 t and had its bucket capacity increased to 107 m³ (140 cu yd).

Other B-E stripping shovels followed: the 76 m³ (100 cu yd) 1850-B in 1963 and the last B-E model the 99 m³ (130 cu yd) 1950-B in 1965: the 1950-B was the only B-E machine to have knee-action crowd, and one of these machines, the 'Gem of Egypt', which commenced work in

Egypt Valley, Ohio, in 1969, was the last of this type of excavator built by B-E.

The supremacy of the two 3850Bs was soon eclipsed. In 1965 a Marion 6360 started stripping overburden at the Captain Mine of Southwestern Illinois Coal Corporation, near Percy, Illinois; this unit remains the largest shovel in the world, as two were intended but only one built. This enormous machine weighs 12,631 t and has a 138 m³ (180 cu yd) bucket on a 65.5 m (215 ft) boom. The 6360 soars to a height equivalent to a twenty-one storey building; eight of its electric motors, these alone totalling 5,960 kW (8,000 HP), drive two hoist drums winding four 8.9 cm (3½ in) diameter cables. Four pairs of crawlers, individual crawlers measuring 13.7 m (45 ft) in length by 4.9 m (16 ft) in height and composed of thirty-six 3 m (10 ft) wide pads, are each supported by a hydraulic jack with a 1.7 m (5 ft 6 in) bore!

In the 1960s B-E proposed stripping shovels, larger than the 6360, the 153 or 168 m³ (200 or 220 cu yd) 4850-B and the 191 m³ (250 cu yd) 4950-B, but neither design resulted in an actual machine.

The last stripping shovel to be built in North America was a Marion 5900, in 1971. This machine, working at Campbell Hill, Illinois, has a unique variable-pitch 80 m³ (105 cu yd) bucket which removes both overburden and a 7 m (23 ft) parting between coal seams; it could well have been a fore-runner of designs for future stripping shovels had this type of excavator continued to be produced and developed. Marion built fifty-three units of the eight models of stripping shovels, introduced in the thirty-one years following the first 5561 of 1940; only two left North America, a 5323 to Brazil in 1954 and the last of nine of that model to the UK in 1961.

A limited number of small stripping shovels

Under construction on the Northamptonshire iron ore field this 5323 was one of only two units of models of Marion stripping shovels introduced since 1940 to leave North America.

One of a small number of Soviet-built EBG stripping shovels.

have been built, probably in the 1950s or 60s, in the USSR. In 1981, for example, there were believed to be only five such machines with capacities of over 15 m³ (20 cu yd) at coal mines in the USSR. The largest appears to be the 35 m³ (46 cu yd) EBG-35/65M which was tested at the Berezovsk coal mine and which is capable of being a 40 m³ (52 cu yd) EBG-40/60 version. There is also an EBG-15/40 with its EBG-10/50 version. Very large Soviet machines have been designed up to 125 m³ (163 cu yd) capacity, and including NKMZ's EBG-100/70, but none appears to have been built, presumably overtaken by large dragline development.

It was the introduction by Bucyrus of the 120-B in 1925 which heralded the end of the railroad shovel as a mining tool – B-E produced its last unit in 1929 – and the birth of a specialised hard-digging machine, the electric mining shovel. The full slew, 3 m³ (4 cu yd), 120-B was steam or electric powered and, mounted on a single pair of crawlers, was highly mobile. The 120-B was replaced only in 1950 by the 150-B/R-B and at the same time the smaller 100-B – which first appeared in 1926 – was replaced by the 110-B/R-B. Also in the early 1950s the 5 m³ (6½ cu yd) 170-B, dating from 1929, gave way to the 6.1 m³ (8 cu yd) 190-B. B-E continued to serve the mining market with the introduction of the larger 280-B in 1962, 195-B in 1968, and 295-B in 1972.

Marion had brought out its 151-M as early as 1945 and introduced the 191-M six years later; both are current models. In 1962–63 the only Marion 291-Ms were made, one unit each for Peabody's Sinclair mine, Kentucky and Lynville mine, Indiana. These had 11.5 m³ (15 cu yd) buckets on 27.4 m (90 ft) booms but the model was capable of handling a 19.1 m³ (25 cu yd) bucket on a 19.8 m (65 ft) boom.

P & H rapidly developed into the leading electric mining shovel manufacturer, in terms of units made, since the introduction of the 1400 in 1944; despite being a relative newcomer to the excavator field compared to B-E and Marion. One hundred and fifty 1400s were made up to 1976, plus twenty-eight DE versions in 1947–73. With Magnetorque, and weighing 172 t and 168 t respectively, these models of up to 6.1 m³ (8 cu yd) capacity are now only made by Kobe in Japan. Although the 1600 was introduced in 1953 it was not until the first 2800s, with 19.1 m³ (25 cu yd) buckets, were erected in 1968–69 that a range of large machines appeared.

As well as B-E, Marion and P & H only China, Romania and the USSR (with three models) produce electric mining shovels over 200 t in weight. The three American manufacturers account for a total of sixteen basic models (excluding the atypical 204-M Superfront) ranging up to the 1,667 t P & H 5700. They also make large crawler dragline derivatives.

The feature which has been of most significance in the development of electric mining shovels within the last decade or so – other than increasing size – is the growth of alternatives to standard Ward-Leonard systems of control and drive. P & H, which was a manufacturer of electrical products before it became a member of the excavator industry, manufactures its own electrical components; unlike B-E and Marion which use outside suppliers such as General Electric of America and Westinghouse. For many years P & H excavators have incorporated 'Magnetorque' electric clutches for hoist operations. This device works on the principle of creating a magnetic attraction between inner and outer members of the clutch. Torque is then transmitted with the required characteristics – high at low speed/high loading and low at high speed/low loading – with the added ability of the device to absorb shock loading.

In the late 1960s P & H developed Electrotorque control to convert AC input to DC operating power without the need for continuously rotating AC induction motors and DC generators. Use of the thyristor, a solid state silicon controlled rectifier, provides conversion and control of excitation power for generator fields by electronics, and full AC power conversion with direct controlled delivery of DC power to drive motors. Regenerative power is produced when braking.

The Statitrol II system of static power conversion and control used by Marion in conjunction with DC drive motors is based on the same principles as Electrotorque. The latest designs used by Marion incorporate microprocessor electronics to reduce operational elements, forced air cooling instead of cumbersome liquid cooling, and plug-in printed circuit boards which facilitate maintenance.

In 1980, on its first 395-B shovel, B-E introduced the Acutrol system of static power conversion and control to AC drive motors. So far two other B-E models of mining shovel have been fitted with this system; by the end of 1983 a total of fourteen Acutrol equipped 395-Bs, 295-B IIs

Marion 191-M with static control at Muskingum Mine, Ohio, in 1983.

and 290-B IIs were in operation. For excavator application a two stage AC to DC and DC to AC system of conversion is required to provide adjustable voltage and frequency to control AC motors. These squirrel-cage induction motors (without commutators or carbon brushes) are operated by controlled frequency current through switches connected to external, solid state, electronic circuits automatically sensing, and instantaneously supplying, the correct amount of power required for load variations. Regenerative power is used for braking, even in an emergency during loss of outside current.

B-E's line of electric mining shovels, in excess of 200 t, have single tubular dipper arms, rope crowd and the smaller ones have two-piece booms. The 150-RB at 214 t has shovel bucket capacities up to 10.7 m³ (14 cu yd) and uses the Ward-Leonard system of power control: it is also available as a dragline. From 1957 to date 127 150-RBs have been made. The 155-B1 at 258 t has capacities up to 14.5 m³ (19 cu yd) and was originally introduced in 1975. In 1968 the 195-B appeared and was upgraded to the B-1 version in 1980. This weighs 343 t and has a 9.9–21.4 m³ (13–28 cu yd) capacity range; forty-two have been made by R-B since 1974 – two made in 1976 were for the vast Itaipu hydro-electric scheme in Brazil. Until recently a dragline version, the

200-B, was offered with a standard bucket capacity of 8.4 m³ (11 cu yd) on a 30.5 m (100 ft) boom.

The 290-B II at 464 t was introduced in 1982 when Acutrol was fitted; it replaced the B I version which had in turn in 1980 replaced the original model introduced in 1977. Over fifty 295-Bs and B I versions were made between 1972 and 1982 before the B II Acutrol version appeared. The 295-B II weighs 667 t with 15.3–34.4 m³ (20–45 cu yd) bucket capacities. A fleet of eleven 295-B shovels, including 30.6 (40 cu yd) machines on coal loading duties, operate at Amax Coal's twin Belle Ayr and Eagle Butte mines near Gillette, Wyoming. Two others at the Santa Rita copper mine, New Mexico, provide

the latest means of excavation in an area which, over nearly three hundred years, has seen Apache Indians and Spaniards with pick and shovel, and steam shovels.

The first of B-E's largest model of mining shovel, the 395-B with Acutrol, started work in 1980 at another copper mine, Twin Buttes near Tucson, Arizona. This 838 t machine has bucket capacities of 19.1–45.9 m³ (25–60 cu yd); features include a box beam cross-tie axle to reduce flexing of the undercarriage frame, a crawler design incorporating new bogie type rollers and

The first 838 t B-E 395-B with AC drive went to work at a copper mine in Arizona in 1980.

plate-disc air clutches allowing steering on the move. Subsequent units have been sold to operators in Australia, Yugoslavia and Canada.

The B-E 300-D, uniquely amongst current draglines, is mounted on four crawlers and weighs 384 t. It is related to the 380-W walking dragline, not a shovel, and was introduced in 1980. Only one has been put to work so far with a 10.7 m³ (14 cu yd) bucket, originally located in Illinois.

Marion's 151-M is available in standard form at 214 t with capacities from 5.3 to 9.1 m³ (7–12 cu yd) and in a long range version of 232 t with the same bucket capacity range. In 1983 eleven units were at work for Amax Coal at the company's Chinook (three machines), Minnehaha (two) and Wright (one) mines in Indiana, and at the Delta (two), Leahy (two) and Sunspot (one) mines in Illinois.

Majdanpek copper mine in Yugoslavia was the destination for the first 182-M introduced in 1966: subsequently three more 182-Ms went to this mine. The current 182-M weighs 323 t and can take buckets up to 13.8 m³ (18 cu yd). During the 1970s 182-Ms went, for example, to the UK

Bucyrus-Erie's 300-D has four crawlers, a unique feature on modern draglines.

and Indian coal mines, and to a South African copper mine. The dragline version of this machine is the 6.1–7.7 m³ (8–10 cu yd) 184-M weighing up to 375 t and with boom lengths up to 45.7 m (150 ft). The 184-M is diesel-electric powered and uses one twelve cylinder and one six cylinder Cummins engine, or two Caterpillar engines.

The Marion 191-M is available in standard form at 463 t with 9.1–22.9 m³ (12–30 cu yd) bucket capacities or in a long range version at 503 t with 7.6–16.8 m³ (10–22 cu yd) capacities. The hundredth 191-M went in 1977 to dig taconite at the Empire Mine on the Marquette iron ore range in Michigan's upper peninsula; this joined other 191-Ms already at work on the nearby Mesabi range in northern Minnesota. Eleven 11.5 m³ (15 cu yd) machines were employed during 1968–76 on construction work associated with the Tarbela Dam Project on the Indus River in Pakistan. Some 122 million m³ (160 million cu yd) of earth and rock were then needed in creating a 3.2 km (2 miles) long barrier to provide water to irrigate 1.6 million hectares (4 million acres) of land and also generate 2.1 million kW (2.8 million HP) of electricity. To the west of the Indus the Parthans had a chief, Baba Khan, over ninety years of age. The third 191-M had 'Baba Khan' painted on its side and a celebration included the blessing of the shovel by the chief and a feast. Guess which one of the eleven machines gave the best production!

A particularly impressive 191-M commenced work in 1966 at the Mount Tom Price iron ore mine in the Pilbara area of Western Australia. This had a 9.2 m³ (12 cu yd) bucket and was powered by three 521 kW (700 HP) Cummins V-12 diesel engines. The crawler dragline version of the 191-M is the 195-M introduced in 1970, the first of which went to OCP phosphate mines in Morocco. Weights range from 524 to 572 t and with bucket capacities of 9.9–13 m³ (13–17 cu yd) it can carry booms up to 51.8 m (170 ft).

The largest of Marion's conventional mining shovels is the 201-M dating from 1975; this differs from the other three basic models in that it has a twin boom and single dipper arm. DC static power conversion and control, rack and pinion crowd, the elimination of oil pumps in the swing gear case, and lower-frame propel are standard features. Bucket capacities for the 613 t standard shovel are 13.8–30.5 m³ (18–40 cu yd) and for the 658 t long range version are 11.4–26.7 m³ (15–35 cu yd). 201-Ms are already working in Canadian iron ore mines, Philippine copper mines and US, Australian and Turkish coal mines. In 1982 a total of twenty-four units were ordered, to be built in the USA by Marion itself and by Sumitomo in Japan, for shipment to countries including Turkey and the USSR.

Six basic models of electric mining shovel make up P & H's current line. These date back to the introduction of the 1600 in 1953; in the following thirty years nearly 800 units of current models and their predecessors were manufactured, a third of these by Kobe Steel in Japan. Australia alone has about eighty-seven P & H electric mining shovels out of a total of some 125 cable machines of this type over 7.6 m³ (10 cu yd) capacity in the country.

With weights up to 254 t, the 1600 is the

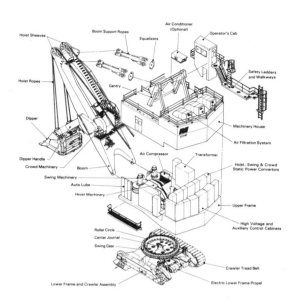

Typical electric mining shovel construction is shown on a Marion 201-M.

smallest P & H machine built in the USA, and has a 4.6 m³ (6 cu yd) standard bucket; a CL coal loader version of 10.7 m³ (14 cu yd) and a CL/LR long range coal loader of 9.2 m³ (12 cu yd) are also available. All 1600s have Magnetorque. The first three of a present total of some fourteen 1600s in the Philippines were delivered in 1968 to a copper mine.

Four versions of the 1900 are available: the 1900 AL, introduced in 1972, is an 11.5 m³ (15 cu yd) standard shovel; the AL/CL a 19.1 m³ (25 cu yd) coal loader; the AL/CL-MR and the AL/CL-LR are medium and long range coal loaders respectively with bucket capacities of 13.8 m³ (18 cu yd). 1900 models weigh around 419 t and use Magnetorque. There are at least seven 1900s in the Philippines: one at a coal mine near Ankara, Turkey; another at a copper mine at Lubumbashi, Zaire; and others at UK coal mines.

The four largest models in P & H's range all now have Electrotorque fitted. The original version of the 2100-BLE was introduced in 1971 and this is the most popular model with 242 made up to the end of 1983. It is a 15.3 m³ (20 cu yd) machine weighing up to some 500 t. 2100B and BLs are amongst eight P & H shovels working at the Panguna mine on Bougainville Island in the

Opposite above : Erected in 1983, this Marion 201-M operates at a coal mine at Cutler, Illinois.

Opposite below : The P & H 2300 was introduced in 1972 and has been in XP version since 1981.

Right : Units of the Soviet EKG-81 have been exported to East Germany, Romania, Turkey and India.

Below : The last model to be made of a line of excavators dating back to 1928 was the B version of the Lima 2400 ; a machine seen in its original distinctive livery in 1968.

Solomon Islands of Papua New Guinea. Here, since 1972, the Rio Tinto Zinc Group has been exploiting a large deposit of copper ore necessitating construction of two new towns, a port, a power station and roads.

Originally brought out in 1972, the 2300 has been in its present XP version since 1981. Weights go up to 731 t with the standard bucket being of 20.6 m³ (27 cu yd) and a long range version of 14.5 m³ (19 cu yd). Both machines can have single or dual motor propel. A 2300 works at Leigh Creek in South Australia and 2300 shovels were ordered in 1982 for use at Soviet coal mines.

The initial 2355 crawler dragline named 'Mr Charlie' was put to work in 1981 in Alabama; by mid-1984 eight had been sold. Modular construction is employed and it can be electric or diesel-electric powered. The 2355 weighs 644 t, has bucket capacities up to 13.8 m³ (18 cu yd), booms up to 61 m (200 ft) and a maximum digging depth of 42.7 m (140 ft).

P & H's 2800XP, in this version since 1982, weighs up to 907 t, has a 27.5 m³ (36 cu yd) bucket and is equipped with either single or dual motor propel. The first 2800s to be made, of 19.1 m³ (25 cu yd) capacity, were also the first to be fitted with Electrotorque. These four units were erected in 1968–69 at the Sparwood coal mine near the Crows Nest Pass in British Columbia. A 22.9 m³ (30 cu yd) 2800 was added later, together with four 11.5 m³ (15 cu yd) 2100s. This mine also operates the world's largest dump truck, the 356 t Terex Titan.

By the end of 1983 a total of sixty-one 2800s had been manufactured, including XP versions. Five 2800s operate at the Black Thunder mine near Gillette, Wyoming – one of the largest coal mines in North America. However, the greatest concentration of P & H electric mining shovels is in Australia at the Mount Whaleback mine. To obtain annually up to 40 Mt of iron ore, from the 5.5 km (3.4 miles) ridge rising above the plains of the Pilbara region of Western Australia, no less than twenty P & H machines (eight 1900s, eight 2100s and four 2800s) are employed – the majority being built by Kobe Steel. Trains carrying 14,000 t each of iron ore, for internal use and export, travel the 426 km (265 miles) to Port Hedland on the Indian Ocean along a railway built in 1967–69.

Largest of all electric mining shovels is the P & H 5700. Mounted on two 13.9 m (45 ft 7 in) long crawlers, 2.5 m (8 ft 4 in) wide, this weighs 1,667 t. As a hard rock shovel it is fitted with a 45.9 m³ (60 cu yd) bucket, and as an LR long range stripping shovel with a 19.1 m³ (25 cu yd) bucket. Like all P & H models the 5700 has a single boom and twin dipper arms. The most powerful version, the hard rock shovel, has electric motors totalling over 4,470 kW (6,000 HP). Two units of this model have been made. A 5700LR began work in 1978 at the Captain mine which, as this joined Marion's 6360, now has both the largest stripping shovel and largest mining shovel in the world on site. A 5700 was commissioned in 1981 at Bloomfield Collieries' mine in the Hunter Valley of New South Wales, Australia.

The USSR has probably only three models of excavator comparable to B-E, Marion and P & H machines in this class. The first Soviet machine capable of handling a 7.6 m³ (10 cu yd) bucket was the EKG-8 which may well have been developed in the late 1960s. The current 8–10 m³ (10½–13 cu yd) EKG-8I model weighs 377 t, has a twin two-piece boom and tubular dipper arm with rope crowd. There is a partly automatic 4 m³ (5¼ cu yd) EKG-4U long-range stripping version, made at the Izhorsk factory in Leningrad, which weighs 366 t and on which the maximum cutting height has been increased by 8.8 m (28 ft 10 in) to 22 m (72 ft 2 in). After tests in 1975 at the Kedrovskiy and Chernigovskiy mines the idea of possible future use of hydraulic crowd was dropped. EKG-8Is have been exported to East Germany and Romania and for use in the coal mining industries of Turkey and India.

Leningrad's Izhorsk factory has also made the 688 t EKG-12.5 since the beginning of the 1970s. This is an unusual machine in that it has four 13 m (42 ft 8 in) full length crawlers in pairs on each side of the undercarriage; each pair is driven separately. The EKG-12.5 has a twin 18 m (59 ft) boom with 13.6 m (44 ft 7 in) tubular dipper arm and rope crowd; bucket capacities are 12.5–16 m³ (16¾–21 cu yd). There is a long range version, the EKG-6.3U with a 6.3 m³ (8¼ cu yd) bucket. By 1981 eighteen EKG-12.5s were at work on Soviet coalfields.

In about 1980 the first two units of the USSR's

at major new projects such as the Neryugrin-skoye mine on the south Yakutsk coalfield, in eastern Siberia, and probably also the Udok-anskrye ore mine.

The WD1200 is made in China, and the ESE8001 in Romania; both have bucket capacities up to 12 m³ (15¾ cu yd). China also produces the 10 m³ (13 cu yd) WK-10.

There are four large crawler draglines in this class of machine not produced by B-E, Marion or P & H and not derived from shovels. The Manitowoc 6400 introduced in 1977 weighs some 502 t and has bucket capacities up to 11.5 m³ (15 cu yd) and boom lengths up to 61 m (200 ft). An 1192 kW (1600 HP) V-16 Cummins diesel engine provides power for the digging and slewing motions using Manitowoc's 'Vicon' – variable independent control – system of stepless variable power transmission, and a six cylinder Cummins engine of 335 kW (450 HP) mounted on the crawler frame provides power for independent hydraulic motors on each crawler.

The Weserhütte SW 760 weighs 270 t and has a bucket capacity range of 5.4–8.4 m³ (7–11 cu yd) with boom lengths up to 57 m (187 ft); in 1984 the 480 t SW1220 was introduced with a bucket capacity of 12.2 m³ (16 cu yd). The American 12220 weighs some 315 t when equipped with a 7.6 m³ (10 cu yd) bucket on a 42.7 m (140 ft) boom; bucket capacities range from 4.6 to 11.5 m³ (6–15 cu yd).

largest model of electric mining shovel, the EKG-20, were put on test. The EKG-20 weighs some 800 t and has a 20 m³ (26 cu yd) standard bucket on a single dipper arm, with a crowd motor attached to the boom. A 'TPD' thyristor-electric DC drive system is fitted. This model was produced by UZTM in response to the need for larger and more powerful machines to operate

CHAPTER 5
Walking Draglines

With individual weights up to some 12 Mt, walking draglines form a major group of the largest self-propelled land machines on earth. Yet these highly specialised and gigantic fabrications, so large and heavy that when they first walk after erection relief of internal stress is often achieved by the shearing of thick steel plates, are controlled by one driver who can place a bucket with poise and accuracy that would do justice to a ballet dancer.

Only nine factories in five countries – the USA, UK, USSR, India and China – make these machines although often part-fabrication is undertaken by sub-contractors in other countries such as South Africa and Australia where walking draglines are widely used. Of course, full erection can only take place at the site of operation.

As with other types of excavator, walking draglines utilize to the full modern technology. Computer-aided design and manufacturing methods are employed and on-board computer systems (such as General Electrics 'Digmate', McDonnell Douglas' 'Digs' and Westinghouse's 'LineBoss'), for example, monitor performance

The first two-line dragline; Page's first machine excavating Lock 27 foundations on the Hennepin Canal, Colona, Illinois, in 1904.

and provide warnings of unsafe operational situations.

Although the Osgood Dredge Company of Albany, New York, had designed and patented in 1880 a dragline excavator, a 'No 15 Steam Shovel to Work Backwards', whether or not any of these were built and put to work is uncertain. However, in 1904 John W. Page of the Chicago-based contracting firm of Page & Shnable devised what has generally been accepted as the first practical dragline with a specialised two-line bucket with in-built dumping facility. The Osgood machine and, for example, the Heyworth-Newman dragline devised by fellow contractors in 1908, were both three-line machines requiring an extra hoist line. The original Page dragline was created when Page & Shnable was building lock foundations on the Hennepin Canal at Colona, Illinois. A combination stiff leg and guy derrick with an 18.3 m (60 ft) boom, operated by a two-drum steam engine, had a wheel-scraper pan shaped bucket attached. Dumping was achieved, without the necessity of the bucket being unhooked manually, by a device known as the Page hitch which became the standard dump line in use ever since. In 1905 the pulling hitch was re-located well forward and above the cutting edge and the bucket shape was changed to one resembling that of the present day.

Page & Shnable not only used this system of earthmoving on its own contracts but also sold dragline buckets to other contractors. The gear ratio on standard steam engines, and vertical boilers, were not suited to dragline use so in 1907 when Page & Shnable was preparing foundations for a steel mill in Gary, Indiana, the company changed to horizontal boilers and ordered special heavy-duty equipment including steam engines with compound gear drums. This equipment came from another Chicago company, the Monighan Machine Works, which had been created around 1884 by John Monighan – a machinist with experience of railway equipment – and which specialised in hoisting machinery. Following the demand created by the success of the Page-Monighan machine at Gary the two companies informally agreed to manufacture draglines jointly; Monighan supplied the machinery, Page supplied his patented buckets and ordered the structural steel. Either company accepted

orders, and the draglines were erected on site. By 1910 five models were available ranging from 0.76 to 1.9 m³ (1–2½ cu yd), and in that year Page-Monighan delivered what was probably the first excavator of any kind to be powered by an internal combustion engine. A dragline with a 37.25 kW (50 HP) Otto petrol engine was shipped to the Mulgrew-Boyce Company of Dubuque, Iowa.

When the Monighan Machine Company was incorporated in 1908 Oscar J. Martinson became Secretary. In 1889, as a nine-year-old boy, he had emigrated from Norway to Chicago and was working as a draftsman when he joined Monighan. Although Oscar Martinson had had limited formal education he rapidly became a practical and inventive engineer and in 1910 patented a two-line dragline bucket that was offered as an alternative to the Page bucket. Only three years later he patented the first excavator 'walking' device against which the traditional mountings for draglines – skids and rollers, or railway wheels running on movable sections of track – were no match. Thus, early in the twentieth century, not only were the foundations laid for two of the mere five present day designers and manufacturers (Page and Bucyrus-Erie/ Ruston-Bucyrus which later took over Monighan), but also these two companies had already pioneered key elements of the walking dragline: the automatic two-line bucket and a walking mechanism.

The 'Martinson Tractor' walking device consisted of a pair of pontoon-like 'shoes' suspended by chains from beams mounted on cams on each side of the machinery house. The two cams were attached to a shaft running through this structure which itself fully slewed on a flat circular 'tub'. When the cams revolved the shoes dropped to the ground and, as the cams rotated along the surface of the shoes, they lifted the dragline's rear and slid the whole machine along the ground suspended on the cams and supported by the tub's front edge. When the cams completed their cycle the upper structure and tub were lowered and the shoes lifted ready for the commencement of another cycle. A dragline so equipped could, therefore, move in any direction; it exerted relatively little ground pressure when digging and could quickly travel long distances.

Walking draglines started with Martinson Tractor device patented by Oscar Martinson in 1913.

Although a few complete draglines had been assembled by Monighan alone as early as 1909, the development of the walking device provided a clear incentive for the company to become an independent manufacturer. Page continued to buy Monighan machinery until about 1916. However, in 1913 Monighan's first walking dragline, with a 2.3 m³ (3 cu yd) bucket on a 12.2 m (40 ft) boom, was delivered to Reeve Hutchinson of Wheeling, Illinois; in the same year the first series produced model appeared, the 3-T with a 2.3 m³ (3 cu yd) bucket on a 18.3 m (60 ft) boom, of which sixty-five were produced up to 1926; and in 1914 the 1-T was introduced with a 0.76 m³ (1 cu yd) bucket on a 12.2 m (40 ft) boom of which 117 were made up to 1925.

In 1926 Oscar Martinson improved his walking device by patenting an enclosed rolling cam mechanism and this basic design is still used on smaller models of walking draglines by Bucyrus-Erie, the company to which Monighan became

affiliated in 1931, and finally absorbed by in the mid-1940s. By 1931 Monighan was producing machines in size up to the 221 t 6-W capable of handling a 4.6 m³ (6 cu yd) bucket; and in the 1920s walking draglines had been used on drainage projects in Arkansas and on levée building along the Mississippi River.

New Bucyrus-Monighan models were developed in the 1930s and 1940s and introduced to additional types of application; first to Pennsylvanian anthracite mining, then to operations on the bituminous coalfields of Indiana, Illinois, Kentucky and Ohio, and later to phosphate workings in Florida. Ruston-Bucyrus built its first walking dragline in 1939; fifty-three US-designed 5-Ws were built by R-B during 1939–71 and twelve 3-Ws during 1941–54.

The 19.1 m³ (25 cu yd) 1150-B, evolved from the 950-B stripping shovel, was first put to work in 1944: five of these models went to the UK to work at coal and iron ore mines. Similarly, the 650-B was derived from the 550-B shovel. Technical developments, by then on 'Bucyrus-Erie' walking draglines, included the development of an automatically lubricated cam and slide walking mechanism with anti-friction roller bearings to replace the Martinson-derived cam and frame device on large machines (1963); the introduction of lighter triangular booms (1964); the fitting of an electrical timing device to eliminate the need for a shaft connecting the walking shoes (1969); and the sharing in 1983 with Marion of the distinction of building the longest dragline boom – the B-E 121.9 m (400 ft) boom on a 2570-W being erected at Poplar River lignite mine, Saskatchewan, and the Marion on a 8750 being erected at Highvale coal mine, Alberta.

Undoubtedly the most significant technical achievement was B-E's production of the world's largest excavator, the 168 m³ (220 cu yd) 4250-W walking dragline – 'Big Muskie' – on which work started in 1966. This mammoth electrically powered machine of 33,524 kW (45,000 HP) commenced operating in 1969 at the Muskingum mine of Central Ohio Coal at Cumberland, near Zanesville, Ohio. The mine supplies some 3 Mt of coal annually to the AEP power station at Beverley, Ohio.

Big Muskie weighs around 12,244 t, equivalent to nearly 130 Boeing 727 aeroplanes or a line of cars some 40 km (25 miles) long! It would fill the width of an eight lane highway; its twin 94.5 m (310 ft) booms are each constructed from four 0.6 m (2 ft) diameter gas filled tubes with interlacing; and it sits while working on a 32 m (105 ft) diameter tub. Walking is achieved by four vertical hydraulic cylinders raising the upperstructure so that four shoes, linked in pairs, bear the machine's weight. The upperstructure is then pushed 4.3 m (14 ft) backwards by four horizontal cylinders before lowering onto the

'Big Muskie's' 168 m³ (220 cu yd) world-leader bucket easily holds a complete band.

World's largest single-bucket excavator, B-E's 4250-W, weighs over 12,000 t; 'Big Muskie' at work in Ohio.

'Big Geordie', largest dragline in western Europe, walks across a road from one coal mine to another in Northumberland in 1973.

tub. Next the shoes are raised, pushed towards the rear of the machine, lowered, and the cycle begins again with a rate of travel of 0.27 kph (0.17 mph).

The detached house-sized bucket digs around 295 t of overburden each time when pulled by 12.7 cm (5 in) diameter steel drag cables, and it can hoist this a total of 99 m (325 ft) – the height of a thirty storey building. Big Muskie moves 23–46 million m³ (30–60 million cu yd) of material annually; working non-stop day and night except for Christmas Day, meal breaks, servicing and repair.

In 1979 the second largest draglines ever built started work; these being two B-E 3270-Ws with 135 m³ (176 cu yd) buckets at coal mines at Chandler, Indiana, and Marion, Illinois.

To date over 800 Monighan and Bucyrus-Erie walking draglines have been made. The smallest

Page 757 at a Montana coal mine – this model weighs 3,426 t and has bucket capacities up to 57.4 m³ (75 cu yd).

A B-E modular 380-W is operating in Oneonta, Alabama.

in the current range of thirteen models is the modular 380-W; the first of which was shipped to Tulsa, Oklahoma in 1979. By the end of 1983 seventeen of these with capacities up to 12.2 m³ (16 cu yd) had been built in the US, and four by R-B in the UK. Two more units were being built by R-B during 1984 for shipping in 1985 to Egypt for phosphate mining. The 480-W is the longest running of B-E's present line, having been introduced in 1955. The first unit of the company's other modular walking dragline, the 680-W with Acutrol AC drive, commenced work in 1983 at Cordova, Alabama.

The 1260-W, introduced in 1964 and of which twenty-four had been built by 1979, has bucket capacities up to 30.58 m³ (40 cu yd); between 1976 and 1980 six were built by R-B, two for shipping to the USA and four for retention in the UK to work at coal mines. The 1300-W first appeared in 1971, the 1350-W in 1967 and the 1360-W in 1976.

1370-Ws, with bucket capacities up to 45.9 m³ (60 cu yd) work at coal mines in New Mexico, Indiana, and the Hunter Valley of New South Wales, Australia, and uncover phosphate in Florida. 1570-Ws, up to 61.2 m³ (80 cu yd), were introduced in 1972. In 1981 one, 'Mr Heman II', travelled 27 km (17 miles) in fifteen days in Alabama, crossing rivers, streams and roads, and accompanied by an ice truck to feed a special cooling system for the walking cam and frame mechanisms. The 1770-W is a recent addition to B-E's line and can take buckets up to 72.6 m³ (95 cu yd) capacity.

The first 2570W with a 76.4 m³ (100 cu yd) bucket began operating in 1972 in Indiana, near Oakland City; one in Alabama is regularly visited by the operating company's three helicopters on mine planning, exploration and sales exercises; and others operate on the Alberta Tar Sands near Fort McMurray. Only the two 3270-Ws and the one 4250-W, described previously, have been produced so far by Bucyrus-Erie.

After the break with Monighan in about 1916 the Page company developed its own range of draglines. In the early years rail and crawler mounted versions were produced but during the 1920s Page devised its own three and four-legged racking systems to 'walk' its draglines. Pre-1940s the '400' series – of which three units were still operational in 1977 in Canada, Mexico and Hawaii – were originally all steam powered: the 430 utilized Page's first walking device, a four-legged system; the 411 was three-legged or crawler mounted, and the 421 had the walking device introduced in 1935 which forms the basis of the present Page system. This consists of two long, vertical welded 'spuds' pivoted to shoes and activated by double cam eccentrics integrated into single castings in each spud. The long throw from one eccentric is used for moving the machine horizontally, the shorter from the other for lifting it vertically. This arrangement causes the shoe to move in an elliptical-shaped path giving a long step but minimum lift.

In 1924 Page designed and built the first diesel engine exclusively for dragline use. Diesel powered machines were subsequently a feature of Page products; the 600 series were all originally diesel powered. Also early models of the 700 series, a series introduced in the mid-1950s of essential electrically powered machines, had models available with Page horizontal-V diesel engines such as the 5.3 m³ (7 cu yd) 721, of 358 kW (480 HP); 7.6 m³ (10 cu yd) 723, of 499 kW (670 HP); and the 9.2 m³ (12 cu yd) 728, of 663 kW (890 HP). In the 1960s diesel engines of up to 1311 kW (1760 HP) were fitted. In the late 1970s the 828 with buckets up to 10.7 m³ (14 cu yd) was available with a Caterpillar V-16 diesel engine of 894 kW (1200 HP), and diesel power was offered on the 732 into the early 1980s.

There are eight models in the present range of Page walking draglines. These have booms of simple T or H chords and angle lacing; directly driven AC electric power systems (created in 1957); and can use any of the wide range of special Page automatic deep-digging buckets (devised in 1930) – including the archless version used for loading hoppers or trucks (introduced in 1960) – with since 1964 the 'Miracle Hitch' designed for fast hoisting.

Five 700 Series machines range from the 897 t 732 with bucket capacities up to 15.3 m³ (20 cu yd) on booms up to 71.3 m (234 ft), through the 736, 740 and 752 to the 3,426 t 757 with capacities up to 57.4 m³ (75 cu yd) on booms up to 106.7 m (350 ft). The three 800 series models, introduced within the last decade, include the 840 – up to 29.1 m³ (38 cu yd), 852 – up to 34.4 m³

This 111 m³ (145 cu yd) Marion 8900 was shipped to the Dugger Mine, Indiana, in 1967, following a similar machine which went to Australia the previous year.

(45 cu yd) and long-range version of the 852 – up to 47.4 m³ (62 cu yd).

Between 1940 and 1977 around 200 Page walking draglines were produced; about 170 of these were working at the end of that period, mostly in North America, but also in Turkey, Australia, Greece, Poland, Pakistan and India. The two 734s introduced into Indian coal mines at South Balanda and Kurasia in 1961–62 were the first walking draglines in that country. Two 752s were put to work in 1976–77 at Balm, Florida, and three 752LRs at Craig, Colorado, between 1977 and 1979. An 852LR and an 852 went to Fort Meade, Florida, in 1980–81, and in 1983 a 757 began work for Union Oil Company at Hinton, Alberta, near Luscar Sterco's Coal Valley mine where a 752LR has been operating since 1977.

Marion produced its first walking dragline, a 7200 electric model, in 1939. In the ensuing forty-four years a further nineteen models were introduced and a total of nearly 300 machines built.

The diesel version of the 7200 came in 1940 and a total of fifty-seven units of both versions were made up to 1958; four still work in the Yugoslavian coal mines at Banovici and Djerdevik. Also in 1940 the 7400 electric version was introduced and with a total of ninety-two of these and (from 1949) the diesel version made up to 1974, the 7400 became the most popular of all Marion's walking draglines. The 7800 was made between 1942 and 1964; one with a 17.6 m³ (23 cu yd) bucket is at the Togston coal mine, Northumberland.

There was a gap of twenty years before the next walking dragline model was introduced by Marion – the 7900 in 1962. In the following year came the first of two 8700s shipped in 1963–65 to Ohio and Pennsylvania. Also in 1963 the only 8800 to be made, which had a 65 m³ (85 cu yd) bucket on a 83.8 m (275 ft) boom – then easily the world's largest walking dragline at some two and a half times larger than any previous machine – was shipped to the Homestead coal mine at Bixby, Kentucky.

In 1965 Marion designed, but never built, its 9600 model. This was in response to Central Ohio Coal's need for either one exceptionally large or two large walking draglines at the Muskingum mine. Although Marion offered the 9600, with a 168 m³ (220 cu yd) bucket on a 91.5 m (300 ft) boom and weighing 12,210 t the company recommended to COC two large machines instead. In the event B-E's alternative design for a single exceptionally large walking dragline was accepted and 'Big Muskie' was constructed.

In 1966–67 the 8800 was upgraded to an 8900; two were made, the first went to Australia, the second to Indiana, equipped with a 111 m³ (145 cu yd) bucket. Two new models were introduced in 1969 comprising a sole 7700 for Australia and the first of three 8400s destined for Morocco, Alabama and British Columbia.

More new models of walking dragline were forthcoming from Marion in 1970 than in any other year. In that year came the 7500 with capacities up to 17 m³ (22 cu yd), and the 7820 up to 34 m³ (45 cu yd)   both of which are current models, together with the 7920 (of which only two were built, during 1970 itself) and the 8000, of which one unit was shipped to South Africa to

Largest of Marion walking draglines, the 8950, operates at Ayrshire Mine, Indiana; a 192-M and two 182-Ms are loading coal.

open that country's first surface coal mine near Hendrina in 1971, and the other shipped to Wyoming in 1971. A 7500 near Somerset, Pennsylvania, had the first production Marion aluminium boom installed on it in 1976.

From 1971 onwards all new introductions are current models: the first of which, the 8750 with bucket capacities up to 92 m³ (120 cu yd) appeared in 1971. Of the twenty-one produced so far most were destined for North America; two, for example, operating at the Universal mine, near Terre Haute, Indiana. The most popular of Marion's large walking draglines appeared in 1972. Thirty-one of these 2948 t 8050 models with capacities up to 50 m³ (65 cu yd) had been made by the end of 1983. At least seven 8050s

operate at coal mines in the remote Bowen Basin of central Queensland, Australia; which they share with rabbits, emus and kangaroos! Another four machines are in the process of delivery.

There is only one 8950, which started work in 1973 at Amax Coal's Ayrshire Mine, Chandler, Indiana. This machine has a 115 m³ (150 cu yd) bucket on a 94.5 m (310 ft) boom; the model which can take buckets up to 138 m³ (180 cu yd) and weigh up to 7,711 t is thus Marion's largest walking dragline and the third largest model in the world. As it was joined by a B-E 3270-W in 1979 the mine now operates with the second and third largest examples of such machines.

In 1973 the 8200 was introduced; machines operate in Kentucky, Wyoming, Montana, Alberta, and on the Kansas and Missouri border. One 7620 was built in 1974 for the Gascoyne lignite mine at Bismark, North Dakota.

Marion's most recent models are of modular

construction for speedy assembly. The 7450 was introduced in 1979; four of the six made to date are in Pennsylvania with the others in South Africa and Alberta. The first 7250 was manufactured in 1983. Both can take buckets up to 12 m³ (16 cu yd) but the 7450 weighing up to 680 t is around 180 t heavier than the 7250 and can carry a longer boom. Also, both can be electric or diesel-electric powered.

The present range of nine Marion walking draglines thus comprises five 7000 Series models with weights up to 1,814 t and bucket capacities up to 34 m³ (45 cu yd), together with four 8000 Series models up to 7711 t and capacities up to 138 m³ (180 cu yd). All models have over and under fairleads and deep H-section, four-chord, booms. Tri-Structure boom supports are available on 8000 Series machines with booms over 61 m (200 ft). These supports weigh less than masts and need less maintenance; they were evolved on models between the first 8400 which went to Morocco in 1969 and the first 8050 of 1972 destined for Australia. Each shoe of 8000

The marks left by the tub and walking shoes of this Rapier W1800 can be seen as it travels backwards down hill: it has worked in South Wales since 1961.

Series machines is driven independently by electrically synchronised motors; the 7000 Series employ a cross shaft. Also 8000 Series machines have saddle mounted outboard propel bearings.

Ransomes and Rapier's involvement with walking draglines was initiated in 1937 when two British companies, Naylor-Benson Mining and Stanton Iron, expressed interest in using such excavators, instead of stripping shovels at iron ore mining operations in Northamptonshire. In 1939, the same year that Marion entered the field, Rapier delivered its first machine – an electric W170 with a 2.7 m³ (3½ cu yd) bucket on a 41.1 m (135 ft) boom – to Naylor-Benson's Nassington mine, and later the same year supplied two diesel W80s with capacities up to 1.9 m³ (2½ cu yd) to Stanton Iron's operations at Market Overton and Eaton.

These Rapier machines were fitted with a 'Cameron and Heath' walking mechanism devised by two of the company's engineers and subsequently used on all models. The shoes are connected by ball and socket bearings to legs containing eccentrics on a drive shaft, or on recent models separate drive shafts. Rollers are interspersed between the eccentrics and the legs, which in turn are connected to links pivoted from

A Rapier W150 working in Sweden – twenty-three units were made between 1944 and 1964.

the main structure of the machinery house. By constraining the legs through these links movement of the eccentrics causes the shoes to move in ellipses, and the longitudinal motion imparted results in a standard walking cycle.

No more W170s or W80s were built; the W90 replaced the W80 in 1943 and was in production until 1954 during which time thirteen were manufactured. In 1944 the first of twenty-three W150s was put to work at a coal mine; W150s had bucket capacities up to 5.3 m³ (7 cu yd) on booms up to 45.4 m (149 ft) and were made until 1964.

In 1947 Stewarts and Lloyds ordered a Rapier walking dragline to strip 30 m (100 ft) of overburden at iron mines at Corby, Northamptonshire. A W1400 with new features, including a 85.9 m (282 ft) tubular boom of triangulated cantilever design which was gas filled

under pressure so that cracks could be detected, was delivered in 1948 – in some respects then the largest walking dragline in existence. The machine weighed some 1693 t and its original 15.3 m³ (20 cu yd) bucket was later replaced by one of 16.8 m³ (22 cu yd). Two more W1400s were built, in 1952 and 1957, for the UK iron ore fields; the 1957 unit, named 'Sundew' after the winner of the Grand National Steeplechase of that year, walked 21 km (13 miles) across parts of Leicestershire and Northamptonshire in 1974. In 1952 the W600 was introduced and in 1956 the W300 appeared.

1961 saw the first W1800 start work at the Maesgwyn Cap coal mine, Glyn Neath, in South Wales where it is still located. This has a 30.5 m³ (40 cu yd) bucket on a 75.3 m (247 ft) boom. Two other W1800s went in 1962–63 to UK iron ore mines and the last, after commencing work in 1964 in Italy, was transferred to Saskatchewan in 1973. In the early 1960s two 1350s were delivered to Canada; one is still in Alberta, the other was moved from New Brunswick to Illinois in the early 1970s.

Around two thirds of the fifty-six Rapier walking draglines made between 1939 and 1964 were W90s and W150s and the destinations of these included Finland, South Africa, Australia and Sweden. During this period a W900 was designed but none built.

From 1964 no Rapier walking draglines were manufactured until a new range was introduced in 1976. Initially the W800, W1300, W2000 and W3000 were offered, to which was later added the W1700, the W1000 and the modular W700. The current range of seven models differs from the earlier range in having conventional tubular booms of rectangular lattice constructions supported by pendants from a mast, shoes and eccentrics on independently driven shafts synchronised by Selsyn units – except the W700 where shoe synchronisation is mechanical – and static

excitation for the main generators. No W800, with capacities up to 15.3 m³ (20 cu yd), W1000, up to 20 m³ (26 cu yd), W1300, up to 26.8 m³ (35 cu yd), W1700, up to 29.8 m³ (39 cu yd) or W3000, up to 53.5 m³ (70 cu yd) models have yet been built. Seven W2000s had been shipped by the end of 1983; the first one started operating in 1978 at the Shannon coal mine, Curlsville, Pennsylvania. The model has bucket capacities of 26–34.4 m³ (34–45 cu yd) on booms up to 95.6 m (314 ft). Two more W2000s went to the USA; 'Brilliant Star' has worked in Winston County, Alabama, since 1979 and a third machine was erected in 1981 at Pottsville, Pennsylvania, but has not yet worked as the mine has been for sale. A W2000 has been stripping some 7 million m³ (9 million cu yd) of overburden annually since 1981 to uncover phosphate deposits at Al Hasa,

Raising the 95.6 m (314 ft) boom on a Rapier W2000 in 1981 at the Al Hasa phosphate mine in Jordan.

The ESh-15/90 built by UZTM is one of the most popular of Soviet walking draglines.

140 km (87 miles) south of Amman in Jordan. Two W2000s are at UK coal mines; one since 1982 in West Yorkshire, the other since 1983 in Northumberland.

Following an agreement made in 1979 W2000s are being built in India for work at Indian coal mines. The first of these was erected in 1983 at Jayant on the Singrauli coalfield in the north-east part of the country. Another was due for completion in 1984 and at least six other W2000s are to be built in India.

The W700, electric powered or equipped with two Caterpillar diesel engines, weighs around 600 t and can take buckets up to 13 m³ (17 cu yd). Five had been built by the end of 1983; the first two were diesel versions delivered in 1981 to coal mines in Pennsylvania. In 1983 three electric W700s became operational in the UK; two at the Godkin coal mine near Heanor, Derbyshire, where they work with the world's largest hydraulic excavator, the O & K RH300, and one at a brickworks at Stewartby, Bedfordshire.

It is unlikely that walking draglines were produced in the USSR before the 1950s. Now two factories make them: Uralmashavod (UZTM) which appears to manufacture a wide range of larger models in limited numbers, and Novokramatorsk (NKMZ) where relatively large numbers of a limited range of smaller machines –

including many for export – are produced. Small, probably Soviet-designed, machines are also built in China.

Soviet walking draglines are designated ESh with the first of two numbers referring to bucket capacity in cubic metres and the second the boom length in metres – the same basic machine with various buckets and booms has different designations.

Early Soviet walking draglines were small, including the ESh-4/40, but by the mid-1970s a range of some seven basic models had been established with capacities up to 100 m³ (131 cu yd) and booms up to 100 m (328 ft). An ESh-125/125 was planned but does not appear to have been built. Rapid growth of surface mining for coal in the USSR stimulated production of walking draglines; during 1970–75 fifty-two new ESh-15/90 and ESh-10/60 (with its replacement the ESh-10/70A) units went to the Kuzbas coalfield alone.

The ESh-6/45, which replaced the ESh-5/45, and the ESh-10/70A are both NKMZ products. ESh-6/45s weigh 300 t and have booms created from single tubes which have supports and cable stays; they now have hydraulic walking mechanisms using two hoist and two traction cylinders replacing original cam systems. The annual output of such a machine is around 1.8 million m³ (2.4 million cu yd). The ESh-10/70A is one of the two most common models of walking draglines in the USSR; it weighs 711 t and can dig some 2.8 million m³ (3.7 million cu yd) of material per year. A full trihedral boom and cam walking mechanism have now been replaced by a two-piece boom and hydraulic walking system. A variant is the ESh-13/50.

UZTM's ESh-15/90A is the other most popular Soviet model. This weighs 1645 t and has a boom similar to the ESh-6/45 and a hydraulic walking system. Further UZTM products, the ESh-25/100, ESh-40/85, ESh-65/85 and ESh-80/100 complete the range of basic models. The 5–6000 t ESh-65/85 is a fairly recent introduction with a trihedral boom, hydraulic walking system and thyristor controlled electric drive.

The first ESh-80/100 weighing 10464 t and with a hydraulic walking mechanism incorporat-

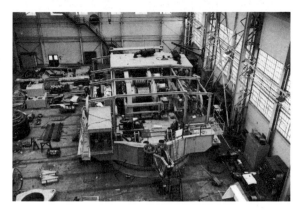

Under construction in 1980 at Ransomes & Rapier's Ipswich, Suffolk, factory was a W700 walking dragline.

ing four shoes linked in pairs, went to work in 1975 at the Nazarovsky coal mine in Siberia. At least one ESh-100/100 version is also operating in Siberia – reportedly capable of excavating with its 100 m³ (130 cu yd) bucket some 20 million m³ (26 million cu yd) of overburden annually.

Soviet-made walking draglines work in many countries outside the USSR; for example, NKMZ machines are in operation in Yugoslavia, East Germany, Poland, Cuba, India and Iran. Two ESh-10/70As were put to work in East Germany, and two ESh-6/45s in Yugoslavia in 1977. The Yugoslavian company, Rudnap, bought an ESh-10/70A in 1978 and a further one with two ESh-6/45s in 1982. One of the 10/70As in Yugoslavia is at the Kostalaz mine. Eight Soviet-built machines were operating at Indian coal mines in 1979 – one ESh-5/45, one ESh-10/60, two ESh-15/90As and four ESh-10/70As.

Other than the recent growth of Soviet manufacturers it appears only two other manufacturers have attempted to enter the walking dragline field since the entry of Marion and Rapier in 1939. In the USA in the late 1970s the American Hoist and Derrick Company promoted specifications for an 1800 Series modular walking dragline, the 1830W, with bucket capacities up to 13.8 m³ (18 cu yd) and booms up to 67 m (220 ft); but none have been made to date. In Romania Promex has worked on the prototype of a 25 m³ (33 cu yd) walking dragline, but it is not known whether or not such a machine has been built.

CHAPTER 6
Construction~size Hydraulic Excavators

The earliest recorded 'hydraulic excavators' were manufactured in the UK by Sir W. G. Armstrong in 1882 for construction of the Alexandra Dock at Hull, Humberside. On these an inclined hydraulic cylinder operated a set of multiplying sheaves to provide the digging action and slewing was achieved by two horizontal cylinders.

A steam shovel, which in many ways reproduced the principles of hydraulic excavators and completely dispensed with cables or chains, was invented in the USA by Frederick O. Kilgore some fifty years before the introduction of fully hydraulic machines. This exceptional type of shovel, available as 30 or 60 t railroad models or mounted on road wheels and with bucket capacities of 0.95–1.9 m³ (1¼–2½ cu yd), had prime digging operations effected by four direct acting steam cylinders. These raised and lowered the boom, pivoted the dipper arm, provided thrust to the dipper arm and swung the mast and boom. A fifth cylinder activated a chain to trip the latch of the bucket door. By 1903 two or three units had been manufactured by the Kilgore Machine Company of Minneapolis, Minnesota, and one was operated by Kettle River Quarries at Sandstone, Minnesota.

In 1914, also in Minnesota at mines of the Penn Iron Mining Company, a unique railroad shovel began work utilising an electric centrifugal pump to operate hydraulic cylinders which activated cables for the hoist and slew, and which had cylinders directly creating the crowd action. Nevertheless, it was the early 1950s before developments took place in western Europe which heralded the beginning of a revolution which was rapidly to transform earthmoving with fully hydraulic excavators.

World War 2 had stimulated progress in the use of hydraulic power and one of the first peacetime applications was to agriculture, where, for example, Harry Ferguson in the UK pioneered the transformation of farm tractors from mere haulage units into versatile mobile power sources. A continent requiring vast reconstruction programmes readily accepted the further development of the tractor into a tool for the construction industry; it first became a loader and then a backhoe loader. Inevitably, despite its many applications, the limitations of such machines were realised and concurrently manufac-

Hydraulic excavators are ideal tools for sewer jobs as shown by this Koehring 1066 excavator doubling as a crane in Massachusetts.

turers began to look at the possibility of creating full-slew hydraulic excavators comparable in size and output to existing cable ones.

Carlo and Mario Bruneri, who had begun making dumpers in 1930, produced in Turin, Italy, as early as 1948 a prototype wheeled machine, used mainly as a front loading shovel and driven by gear pumps and hydraulic motors. After further prototypes, in September 1951 they took out Italian patent No. 483725 for 'a machine the functions of which, 360° slewing and travel included, are fully-hydraulic operated, equipped with a backacter bucket with a capacity superior to 500 litre' (i.e. 0.5 m³, equivalent to $\frac{5}{8}$ cu yd).

The innate caution of contractors in a conservative industry, coupled with a lack of financial resources to adequately promote a new product, resulted in the Bruneri brothers ceding the patent rights to the French company Société Industrielle de Construction d'Appareils Mécano-Hydrauliques (SICAM) in 1954. In the same year SICAM began production at Chaufailles in south-east France of the truck mounted, full slew, Yumbo S25. This was followed two years later by the H25, a self-propelled wheeled machine, weighing some 7.4 t, which in turn was soon joined by the Y35 crawler version.

SICAM's marketing of Yumbo excavators on a world scale later involved licensing arrangements with Drott in the USA, Mitsubishi in Japan, TUSA in Spain and Priestman in the UK. Whether or not TUSA actually made Yumbo machines is unknown, but Priestman in the 1960s only marketed French-made Yumbos. Mitsubishi's licence ran from 1960 to 1977 and the company manufactured its 'Y' series of Yumbo-designed excavators under licence in Japan from 1961 to 1972, starting with the Y35 and H25. Drott obtained the manufacturing rights of Yumbo machines in 1962, but after only two years produced its own design of hydraulic excavator.

The Bruneri Company had reserved the right to manufacture and market the Yumbo in Italy and by 1963 Bruneri itself had built a thousand units. In 1965 the Bruneri brothers dissolved their partnership to lead separate undertakings: the Simit Company (which eventually became Fiatallis in 1974) and the Hydromac Company which operated until closure in 1983.

Poclain's tractor-drawn TO was introduced in 1956.

Across the Alps at the village of Le Plessis-Belleville, 40 km (25 miles) north of Paris, in the same year as the Bruneri brothers established their company, Georges Bataille set up a production workshop to manufacture agricultural vehicles. On the site was a pond ('poche' in the local French dialect pronounced 'poque') used for retting flax ('lin'). The name of this area, 'poque à lin', abbreviated to 'poclain', was adopted by Georges Bataille who called his new company 'Ateliers de Poclain'. In 1948 Poclain made a lorry-mounted loader, and in 1951 brought out its first hydraulic excavator, the TU. This employed high pressure hydraulics and was either mounted on two wheels and towed by, and took its power from, an agricultural tractor, or was mounted on a truck.

The TU was replaced by the TO in 1956 and during that year the clamshell grab was added to the backhoe and shovel already in use. However, it was only in 1961 (the same year O & K brought out its first hydraulic excavator the RH5) that Poclain introduced its first self-propelled and full slew machine, the 10.2 t TY45; over 30,000 of which were eventually sold.

Three West German companies were quick to realise the potential of hydraulic excavators in the 1950s and developed relatively sophisticated machines. Heinrich Weyhausen had founded in

*Cut-away diagram of a Priestman Mustang 120
showing typical elements of a hydraulic excavator:
diesel engine – purple; hydraulic pumps – green; slew
system – red; power controls – yellow; and crawler drive
– blue.*

1919 the company which was to take the name
'Atlas' in 1936; agricultural machinery was produced and in 1949 a hydraulic, tractor-mounted
grab was manufactured. Two years later a truck-
mounted version was added and in 1954 a fully
hydraulic, self-propelled, wheeled machine
emerged from the Delmenhorst works. The
streamlined AB1500, which had a telescoping
boom, could be used as a crane, grab, shovel or
backhoe and went into series production in 1956.

Demag-Deutsche Maschinenfabrik AG – was
founded exactly a century earlier than Atlas and
had made cable excavators since the early 1920s.
In 1954 it too came out with a fully hydraulic
excavator, the middle-pressure 12 t B504
powered by a three cylinder Deutz engine. This
was, however, mounted on crawlers and it had an
HL-shovel version with 0.4–1.5 m³ (½–2 cu yd)

buckets and a TL-backhoe version with
0.3–0.4 m³ (⅜–½ cu yd) buckets.

The third German firm to develop a hydraulic
machine in the 1950s was that founded by Hans
Liebherr who had taken over the family building
business at Kirchdorf in 1938. In 1949 he built
his first tower crane, the beginning of an exceptionally rapid creation of an extensive industrial
empire. In 1957 the first Liebherr excavator,
fully hydraulic from the start, the wheeled L300,
left the Kirchdorf factory.

As might be expected in the USA, the home of
the cable excavator industry since the 1830s, the
application of hydraulic power in innovative
ways to earthmoving was not being neglected.
Nevertheless, concurrent developments here –
partly through the use of low pressure systems
and the creation of hybrid machines retaining
many mechanical features – did not result in
growth on a scale comparable to that in western
Europe.

The basis of what appears to be the first
American hydraulic excavator – albeit an atypical
type – was that devised by Ray Ferwerda and

developed from 1945 by Warner & Swasey into the Gradall telescoping boom excavator. In 1958 W & S also purchased the Badger Machinery Company of Winona, Minnesota, and from their Hopto tractor-mounted backhoe created a range of self-propelled hydraulic excavators.

Bucyrus-Erie acquired the Milwaukee Hydraulics Corporation in 1948; their truck-mounted combined hydraulic and cable H2 Hydrocrane had been invented by R. O. Billings and put into production in 1946. B-E developed this crane and clamshell into the H3 Hydrohoe

Construction of a new board mill at Edson, Alberta, in 1983, with a Drott 45 Cruz-Air.

backhoe in 1951; the basic machine equipped as a crane has been available until recently. It was, though, not until the mid-1960s that B-E manufactured self-propelled hydraulic excavators with the 20-H appearing in 1965, the 15-H in 1966 and the 30-H in 1967.

Other US manufacturers soon joined W & S and B-E during the 1960s as makers of hydraulic excavators. Hein-Werner produced its first hydraulic excavator in 1961 under licence from the Hydraulic Machinery Company of Butler, Wisconsin – but a year later started to manufacture machines to its own designs. The first Drott-built Yumbos also appeared in 1962 and in 1963 Insley entered the market with the H-100 and Koehring with its 505 Skooper. In 1964 Drott began manufacture of its own hydraulic

Twenty-four years of progress: Demag's first model of hydraulic excavator, the B504, and its largest, the first H241 introduced in 1978.

excavator, the Cruz-Air wheeled model; the Schield Bantam 450 appeared – by then the company was part of the Koehring organisation; and P & H acquired the hydraulic backhoe recently developed by the Cabot Corporation's Machinery Division at Pampa, Texas. The following year Unit's H-201-C was introduced. In 1967 Link-Belt produced a range of hydraulic excavators; in 1968 the first Deere machine, the 690, was manufactured; and in 1969 Northwest entered the field.

Both the UK and Japanese industries were in turn founded on licensing arrangements from US, French or West German hydraulic excavator manufacturers. The first hydraulic excavator built in the UK was the Hymac 480. Rhymney Engineering Company in South Wales began production of the HM480, with 270° slew, in 1962. This was a 'Hy-Hoe' built under a licence from the Hydraulic Machinery Company, and the first of an initial batch of six American-built

machines was actually assembled at Rowsley, Derbyshire. In 1963 Hymac produced, with the help of industrial designer Rossi-Ashton, its own full slew machine, the 580, of which over nine thousand were made in various forms during the ensuing two decades.

J. C. Bamford, another manufacturer with origins in agricultural machinery, introduced in 1965 the JCB 7, its first full slew hydraulic excavator. This was an amended version of the W & S Hopto machine and from it evolved the 7 C in 1966 and the 8 C and 8 D in 1971.

Ruston-Bucyrus used the highly successful cable 10-RB, originally designed by its parent American company, Bucyrus-Erie, in the 1930s, as the basis for its own 10 t 3-RB fully hydraulic excavator. This was made in 1963–64 and incorporated two variable-delivery piston pumps.

A Hymac 480 at Rowsley, Derbyshire, where the first units of this model were assembled.

The first hydraulic excavators to be made in Japan were Mitsubishi-Yumbo machines in 1961, including the Y35.

Manufacture of B-E designed hydraulic excavators from 1967 to 1974 delayed the introduction of further R-B designed ones until series production of the 150 RH commenced as recently as 1976.

In the early 1960s Priestman started design work on the fully hydraulic, wheeled, Mustang 90, but this was not launched until 1967. However, prior to the Mustang 90's appearance Priestman had produced the Hydrocub with partly hydraulic upperworks based on a Cub VI and, in 1964, the Beaver with fully hydraulic upperworks but a mechanical undercarriage. The Beaver with an up-rated bucket capacity and a German Geema undercarriage became the T-Two in about 1968 and was made for a short time immediately prior to the introduction of the crawler mounted Mustang 120 Mk I created from the Mustang 90.

The Koehring (Schield-Bantam) 0.76–1.1 m³ (1–1½ cu yd) 475 backhoe was made as an NCK-Rapier model between 1971 and 1976 and a Newton Chambers designed shovel attachment of 0.86 m³ (1⅛ cu yd) capacity was offered. The Koehring-designed 466 backhoe and shovel and 505 backhoe and Skooper were also made in the late 1960s and early 1970s.

Sometime between 1960 and 1962 Smith produced a prototype hydraulic excavator, the Super 10. This did not go into production and the only hydraulic excavators the company made were French Tractems under licence. Between 1968 and 1973 the wheeled ST804 was in production and the ST805C, a derivative of the ST804 with a Smith designed crawler undercarriage and modified upperworks, was produced from 1970 to 1973.

The Japanese hydraulic excavator industry was launched in 1961 when the first Yumbo machines made by Mitsubishi left the Akashi factory; these were both 0.25 m³ (⅜ cu yd) models, the Y35 at 8.3 t and the H25 at 7.7 t. Over 20,000

Mitsubishi-Yumbo excavators were made before Mitsubishi designed and manufactured its own machines from 1972 onwards. In 1962 Yutani obtained a manufacturing licence from Poclain and the following year an agreement was made between Komatsu, Mitsui and Bucyrus-Erie for Komatsu to produce and market excavators – including eventually the 10, 12, 15 and 20-HT hydraulic models – using B-E technology. O & K granted Japan Steel Works a licence in 1964 and the RH5 was the original model to be made by JSW.

Hitachi's crawler mounted UHO3, introduced in 1965, was the first hydraulic excavator to be both designed and manufactured in Japan; the wheeled WHO3 was added three years later. In 1966 Atlas excavators began to be made under licence in Japan by Kubota. 1967 saw Kato become the second Japanese manufacturer to produce its own designed hydraulic excavators. Also in 1967 production of P & H designed hydraulic machines commenced with the H208 made under licence by Kobe Steel. Possibly in 1967, or soon thereafter, Sumitomo started manufacture of Link-Belt hydraulic excavators under licence.

In the early 1970s Japan created, with its own technology, a new type of hydraulic excavator, the crawler or wheel mounted hydraulic mini-backhoe with weights under 5 t. Yanmar produced its first such machine in 1970, Takeuchi in 1971 and Kubota in 1974. There are now at least ten Japanese manufacturers of mini backhoes and this type of excavator is now made elsewhere; for example, in the Netherlands by Vermeer and in the UK by Manitou.

The growth of construction-size hydraulic excavator manufacture was rapid and widespread. Indicative of this is the fact that in the mid-1950s it was estimated that some 10,000 excavators, all cable, were operating in the UK (excavating some 800 Mt/yr of material). Twenty-five years later the number of hydraulic excavators in the UK was estimated also at 10,000, the vast majority of which were construction-size machines which had replaced a large proportion of the cable ones. Currently around a hundred factories world-wide make construction-size hydraulic excavators, hence no more than a brief review of a limited number of manufacturers, and some of the numerous models and their applications, is possible.

One of the largest present ranges of hydraulic excavators in this class of machine comes from

Excavator manufacturing licences in Japan.

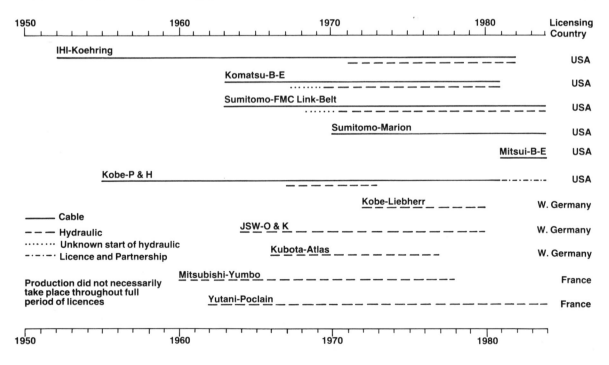

one of the oldest manufacturers, Poclain. Excluding the 600CK (re-numbered 610CK in 1984) and 1000CK there are ten basic models. The smallest, introduced in 1982, is the 5 t 35CK crawler mini model with a 28 kW (37 HP) engine and 0.2 m³ ($\frac{1}{4}$ cu yd) backhoe bucket which can be reversed and used as a shovel. The 60P-B, PL-B and CK-B (first introduced in 1974, 'B' models are recent versions) and 70CK-B (1982) backhoes follow. Larger models can be equipped as backhoe or shovel: 75 P-B and 75 CK-B (1974); 85CK-B (1982); 90 P-B and CK-B (1974); 125 P-B and CK-B (1982); 160CK (1976); 220CK (1976); and 350CK (1982). The 70 t 350CK, largest in this class, has a 277 kW (376 HP) V-10 Deutz or 250 kW (335 HP) six cylinder Cummins diesel engine, or electric motor; bucket capacities go up to 3.2 m³ ($4\frac{1}{4}$ cu yd) as backhoe and 4.4 m³ ($5\frac{3}{4}$ cu yd) as shovel.

In Liechtenstein, the tiny country at the foot of the Alps between Switzerland and Austria, a wheeled 75P operates in and around the village of Triesenberg, digging foundations for houses on steep hillsides. In Western Australia, near Freemantle, a 220 excavated the trench for a pipeline taking sewage out to sea; in Venezuela a 160CL worked on construction of an industrial complex east of Lake Maracaibo; in Sri Lanka, near the island's former capital Kandy, a 220CK has been employed on the Mahaweli River irrigation and hydro-electric power scheme; and recently in the UK in South Wales a variety of Poclain models has been used on building a new 10 km ($6\frac{1}{4}$ mile) road in the Taff Valley.

The O & K range of nine basic construction size machines, RH model numbers 3, 4, 5, 6, 9, 12, 18, 30 and 40, have no less than thirty-eight versions covering wheeled and standard or long crawler mountings, heavy duty versions, and dozer blade and stabiliser options. Weights range from 9.3 to 74 t. The RH30C and RH40C have TriPower shovel equipment.

RH25 and RH40 backhoes were used in 1979–80 along the Bratislava-Komarno section of the River Danube bordering Czechoslovakia and Hungary on a scheme to boost the freight carrying capacity of the river and provide power. Two RH40 shovels and an RH30 backhoe began work in 1981 at a quarry in Libya providing limestone for use at the Mediterranean port of

Homs; while at the same time a couple of RH9s were on one of the islands in the Mediterranean itself helping to create a major water supply scheme on Cyprus. Starting in 1969 no less than twenty-five RH6s dug the trench for a pipeline from Durban to Ogies in South Africa.

All Liebherr's ten current basic backhoe models (excluding the R991) have been introduced since 1980. A wheeled version is denoted by 'A', a crawler version by 'R'. The A900 weighs 12 t; R900 – 14 t; A902 – 13 t; R902 – 16 t; A912 – 17 t; R912 – standard LC or IID up to 21 t; A922 – 20 t; R922 – standard or LC up to 21 t; R932HD – 24 t; A942 – 32 t; R942 – LC or HD up to 31 t; R952 the most recent model, introduced in 1983, weighs 40 t; R962LC – 48 t; R972HD – 64 t; and R982 – LC or HD up to 88 t. The R942, R972 and R982 models are also available as shovels. Six R982HDs, with either shovel buckets of 4.3 m³ ($5\frac{1}{2}$ cu yd) or backhoe buckets of 4.6 m³ (6 cu yd) operated in the early 1980s on jobs associated with the Rio Jequete Peque irrigation scheme bordering the Pacific Ocean in northern Peru; at the other side of the world other R982HDs equipped as backhoes also work on irrigation projects in the fertile lands between the Tigris and Euphrates in Iraq, carrying on a traditional activity dating back to one of the earliest of civilisations which flourished in that area.

Atlas excavators currently range from the 2.8 t AB250 mini to the 52 t AB2502B backhoe, one of which in the mid-1970s elevated and widened the road between Santa Fé and Santiago del Estero to protect it when the River Rio Dolce floods this part of the Argentinian Pampas swamplands. Considerably nearer home territory for Atlas, AB1202, 1302D, 1602D and 1702 models, were responsible for much of the public works required to pedestrianise the main shopping street of Heidelberg in West Germany – famed location of Sigmund Romberg's operetta *The Student Prince*.

In 1983 Bucyrus-Erie had a range of four '100 Series' hydraulic backhoes under 100 t: the 300-H, 325-H, 350-H and 400-H introduced in 1977, 1978, 1975 and 1980 respectively. Weights ranged from 25 to 70 t. These machines had demand/control high pressure hydraulics, independent torque control swing, an MXT propel

system on the 400-H which eliminated chain and track pad bolts for less maintenance, stress flow design and US Steel Corporation's premium grade high stress 50N steel in front end components, and rust-free TriArmor multi-layer fibreglass cab and machinery housings. Another US manufacturer, Koehring, has eight excavators within this category: the current version of the 166 dates from 1982; those of the 266, 366, 566, 666, 866, 1066 and 1166 are all from 1981. The 866, for example, weighs 61 t and is of 336 kW (450 HP); the 1066 weighs 76 t and is of 340 kW (456 HP).

The USSR produces at least eight models of construction-size hydraulic excavators. The smallest, the EO-5015A, sometimes marketed as the 'Belaz', is a crawler backhoe of 56 kW (75 HP) weighing 12 t with bucket capacities of 0.5–0.8 m³ ($\frac{5}{8}$–1 cu yd). During 1972–73 eleven units were sold both to Czechoslovakia and to Hungary, and thirty-seven to Bulgaria. The EO-3322B, at 15 t, is a wheel mounted backhoe or shovel; its 'A' version appeared about 1974 and in 1978 a 'V' version was introduced with

To dig foundations for a Hyatt Regency hotel in Milwaukee, Wisconsin, the contractor used B-E 20-H and 300-H backhoes.

The Roman amphitheatre at Arles in southern France makes an impressive background for this Richier backhoe.

The largest of Caterpillar's range of excavators is the 245 weighing up to 67 t.

semi-automatic parallel digging action. A hundred A versions went to East Germany in 1974, and a further hundred B versions in 1978. The crawler EO-4123 and wheeled EO-4321 are variants of the same basic 0.8 m³ (1 cu yd) backhoe or shovel excavator and weigh 18 t. In 1978 thirty wheeled versions were shipped to East Germany. The EO-4121 was introduced about 1974 and the 'A' version appeared around 1980. It is a 21 t, 97 kW (130 HP) crawler machine with backhoe and shovel ranges of 0.73–1.2 m³ ($1-1\frac{5}{8}$ cu yd); thirty were supplied to Hungary in 1978. In the early 1980s a new model, the EO-4124 appeared, possibly as a replacement for the EO-4121.

Normally equipped as a 1.6–2.8 m³ ($2-3\frac{5}{8}$ cu yd) shovel, but also available as a backhoe, the crawler EO-5122 was introduced in 1975; it weighs 38 t and is of 127 kW (170 HP). Twenty-five EO-5122s with 1.4 m³ ($1\frac{7}{8}$ cu yd) buckets were delivered to Thailand in 1982–83. There is

now an EO-5123 model. The largest Soviet machine is the electric powered crawler EO-6122 first produced in 1978. This, as shovel or backhoe, weighs up to 59 t with bucket capacities up to 5 m³ ($6\frac{1}{2}$ cu yd). Construction-size excavators built in the USSR have variable displacement pumps with automatic power control, and Yaroslavl diesel engines are fitted.

Mitsubishi's line consists of ten basic models of hydraulic backhoe with weights of 2.8–61 t: these are all new 'MS Series' machines created since 1979 embodying energy saving and working life prolongation features. The MS070-2 is a direct descendant, via the MS20 introduced in 1973 and MS062 introduced in 1975, of one of the first hydraulic excavators to be made in Japan – the Y35. Another model, the MS110-5, is descended from the Y55 introduced in 1967, via the MS40 which appeared in 1972. During the first twenty years of production, fifty thousand Mitsubishi excavators were made. The two largest current models, the MS380-2 and MS580 can be used as shovels; the MS070L-2 is a low ground pressure machine, and the boom of the

Opposite above : A contractor in Triesenberg, Liechtenstein, uses a Poclain 75P for house foundation excavations on Alpine foothill slopes.

Opposite below : First Italian Laltesi excavator was the L30, shown here with the prototype L200.

Right : Newly introduced when photographed in 1982 this JCB 807C is at Ryefields coal mine, Derbyshire, with two Lima 2400B draglines and a 22-RB shovel amongst the machines in the background.

Below : Introduced in 1983 the JCB 802 mini backhoe incorporates Kubota components.

EXCAVATORS

MS070U-2 can be swung horizontally to allow offset digging up to 110° to the right and up to 60° to the left.

Since 1979 when Kobe Steel began manufacturing all its hydraulic excavators to the company's own designs, following licensing agreements with P & H (1967–71) and Liebherr (1972–79), around 14,000 Kobelco units have been produced. The present range of eight basic machines in this class extend from the 6.6 t K903B backhoe to the K935 at 59 t as a 2 m³ (2⅝ cu yd) backhoe or at 60 t as a 3.5 m³ (4½ cu yd) shovel.

JCB in the UK now makes four crawler backhoes. The 802 mini, introduced in 1983, weighs 3.3 t; it has bucket capacities up to 0.13 m³ (⅛ cu yd) and a maximum digging depth of 3.1 m (10 ft 2 in). The 805B Turbo with weights up to 13 t, 806C up to 14 t and the 807C up to 21 t were all introduced in 1982; these latter three models have parts of their front end equipment trapezoidal in section, and can be equipped with two piece booms incorporating the Powerslide system of variable geometry which was first applied to the original 805B in 1979.

Åkerman, particularly noted for pioneering environmental and safety features, such as sound insulation, now build a range of five models in Sweden together with some in the USA. These are the 15 t crawler H7BLC and its 14 t wheeled H7MB version; the 21 t crawler H10 and its 19 t wheeled H10M version; the crawler 27 t H14B; the crawler H16D with weights up to 38 t; and the crawler H25C with weights up to 57 t.

Five Åkerman excavators – two H25Cs, an H9B, and two H7MBs, were employed in 1983, by Tarmac the contractors, repairing the M1 Motorway in Northamptonshire. Also in the UK, near Dunbar in south-east Scotland, six Åkerman machines including an H25B, H25C and an H16C, recently worked on the Torness nuclear power station project. The Majes irrigation scheme in Peru, which was initiated in 1974, had two H25Cs digging at 4,100 m (13,400 ft) above sea level in the Andes on work associated with the Condorama Dam; elsewhere H12s were being used on the scheme. Since 1980, up to five H25Cs have been in Thailand at the Khao Laem hydro-electric construction site around the River Quae Noi (River Kwai) on quarry duties for rockfill, foundation works and spillway and diversion channel excavation. Ten Åkerman excavators, including six H25Cs, were ordered in 1983 to assist in building a power station in Java for the Indonesian Government.

CHAPTER 7

Hydraulic Mining Shovels and Backhoes

Whereas the dramatic increase in size of cable excavators during the last forty or so years took place on a long-established base, the equally significant development of large hydraulic mining shovels and backhoes has been achieved from scratch in this same relatively short period of time. There are now twenty-three models of hydraulic excavator in excess of 100 t weight: of these two are around 500 t, three in the 260–270 t range, eight between 150 and 190 t and the remaining ten go up to 144 t. Between 1975 and the end of 1981 some 800 such machines were produced challenging, and at times replacing, the smaller sizes of cable mining excavators.

Although US manufacturers produced hydraulic excavators around 100 t during the 1970s – such as the Northwest 65-DS and 100-DH, the W & S Hopto 1900 and the Koehring 1266D – it was French and German companies which led the field. Japanese, Soviet and larger American machines in this category began to appear in the late 1970s and early 1980s.

Poclain broke into the mining market in 1971 when production commenced of the first fully hydraulic excavator in excess of 100 t. The EC1000, at 137 t, was easily the largest hydraulic excavator with its three GM Detroit 8V71 diesel engines totalling over 596 kW (800 HP) and bucket capacities of 5 m³ (6½ cu yd) as backhoe and 8.8 m³ (11½ cu yd) as shovel. Tried and tested engine and pump modules used on the HC300 since 1968 were incorporated into the EC1000. The first major bulk earthmoving task undertaken with a large hydraulic excavator as prime mover was the construction of part of the Paris-Strasbourg motorway when, early in 1974, an EC1000 shovel was put to work by Razel Brothers. In 1975 the first production unit of the EC1000's replacement, the 162 t 1000CK Series I, went to work at a UK coal mine near Leigh, Greater Manchester; subsequently Series I

machines were used at sites including the Paluel nuclear power station on the French coast at Caux, Normandy, and at a Spanish lignite mine near La Coruna in Galicia.

The current 1000CK Series II model was introduced in 1980 and its weight has been increased to 190 t. The normal diesel version employs two six-cylinder Cummins engines or the machine can be powered by electric motors. Like other Poclain excavators a 'Variodyn' variable flow hydraulic circuit offering independent and simultaneous operation of all functions is fitted. The largest bucket which can be used is a 17 m³ (22 cu yd) coaling one, and an average annual earthmoving production of some five Mt can be achieved. One of the first Series II machines was used to dig hard sandstone at the Lynn Land coal mine, 161 km (100 miles) south of Charleston, West Virginia.

Poclain's other mining excavator is the 600CK (now 610CK) Series II, introduced in 1981 to replace the 107 t Series I model, which dated back to 1975: Series I machines had seen service, for example, on demolition work in Paris, sewer line excavation in Virginia and port construction at Zeebrugge, Belgium. The new 600CK is a 120 t excavator with bucket capacities ranging from 3.1 m³ (4 cu yd) to 5.4 m³ (7 cu yd) as backhoe, and from 5.5 m³ (7¼ cu yd) to 11.5 m³ (15 cu yd) as shovel. Electric power is available, or a choice of two water cooled six cylinder Cummins or two air cooled V-10 Deutz, diesel engines. A 600CK Series II backhoe has recently been at Riemst-Vroenhoven in Belgium widening the Liège-Antwerp Albert Canal and excavating down to a depth of 6 m (20 ft).

In 1978 Demag took over from Poclain the position as manufacturer of the world's largest hydraulic excavator. The original H241 weighed 238 t; now it weighs up to 280 t and can be equipped with bullclam shovel buckets from 10

81

Opposite : An 8.3 m³ (11 cu yd) Poclain 1000CK Series I shovel digging in the late 1970s on the chalk and silt cliffs at Caux, Normandy, for France's Paluel nuclear power station.

Excavators often have to work round the clock; two Poclain 1000CK Series I backhoes at a lignite mine in the Galicia region of north-west Spain.

World's largest hydraulic excavator is the 1,730 kW (2,352 HP) o & K RH300 shown operating at the Godkin coal mine in Derbyshire in 1983.

to 21 m³ (13–27 cu yd) and backhoe buckets from 5 to 21 m³ (6½–27 cu yd). A 152 t truck can be loaded in five or six cycles within two and a half minutes. Power for the diesel version is supplied by a single, water cooled, GM Detroit 16V-149T, turbo-charged, sixteen cylinder two stroke engine; this is set to develop 970 kW (1318 HP). The power take-off unit drives five hydraulic variable displacement axial piston pumps.

By the end of 1983 over sixty H241s had been put into service throughout the world, at least a quarter of these in North America. The Quintette coal mine at Tumbler Ridge, near Prince George, British Columbia, will employ eight H241s by 1986 (together with three P & H 2800 electric mining shovels). This new mine, together with the nearby Bullmoose mine, has necessitated construction of a 130 km (78 miles) branch railway between 1981 and 1983 with bridges and long tunnels, upgrading of the existing main line, and development of a new port south of Prince Rupert which first shipped coal to Japan at the beginning of 1984. An H241 electric backhoe at a coal mine in the Whitbank area of the Transvaal, South Africa, has a 'power pack', consisting of a 300 kW (400 HP) diesel engine, hydraulic pumps and oil tank mounted on a wheeled chassis, that can be hooked behind the machine. Hoses transmit hydraulic power to the H241s drive motors on long-distance moves to restore the mobility an electric-powered excavator would normally lose because its power comes via a trailing cable.

The Demag H185 appeared in 1983 and within the first year of production some twenty units had been sold including three scheduled for work at UK coal mines and six destined for Thailand coal mines. Machines had also been sold to operators in West Germany, USA, South Africa, Spain and Mexico. The H185 is a 175 t machine with shovel and backhoe capacities from 7.5 to 15 m³ (10–20 cu yd).

Demag's third mining excavator, the H121, replaced the H111 in 1979. At about 115 t the H121, has bucket capacities up to 10.5 m³ (14 cu yd).

World leadership as far as size of hydraulic excavators is concerned was soon taken from Demag. In 1979 the H241 was easily surpassed by O & K's RH300 at no less than 498 t. With the possible exception of the Soviet EG-20 – which may be slightly heavier – it remains the largest machine built; although only three have been manufactured and of these one put into service by the end of 1983. The first RH300 equipped as a 22 m³ (29 cu yd) shovel was bought by Northern Strip Mining and started work at the beginning of 1980 at the Donnington Extension coal mine near Swadlincote, Derbyshire. It has since been transferred to another of NSM's operations in the UK, the Godkin mine, near Heanor, also in Derbyshire; where it works in conjunction with three Wabco Haulpak 170C dumptrucks of 173 t payload.

A coaling bucket of 34 m³ (44 cu yd) is possible for use on the RH300, two-piece boom and monoblock backhoes are offered with bucket capacities of up to 29.1 m³ (38 cu yd) and pontoon-mounted dredging versions could dig at depths down to about 24 m (79 ft) below water level. Both the overall height of the machine and the undercarriage width are 7.1 m (23 ft 5 in) and 1.5 m (4 ft 10 in) wide tracks are fitted. Power is from two water cooled Cummins KTA 2300C 1200 diesel engines, supercharged with intercooling. These engines provide a total of 1,730 kW (2,352 HP) which creates both crowd, and penetration – breakout force on shovel teeth, forces of 224 t. Each of the two parallel connected diesel engines, or alternative electric motors, controls four variable displacement axial piston pumps so that all functions can be utilized at half speed, but with full power, with only one engine running. Seven identical cylinders operate the digging equipment.

The standard RH300 is designed to operate in temperatures from + 50°C to − 28°C and up to heights of 3,600 m (11,811 ft).

O & K's RH120C was introduced in 1983; the first production unit equipped as a shovel began work at a landfill project near Peterborough, Cambridgeshire, at the end of that year. Weights go up to approximately 184 t equipped as a shovel with capacities of 8.5–17 m³ (11–22 cu yd) and 182 t as backhoe with capacities of 5.6–14.5 m³ (7¼–19 cu yd). The RH120C has the recently developed O & K 'TriPower' system of parallel kinematics for shovel equipment. The TriPower system provides parallel bucket action when crowding at the face, and lifting to the dump position, by employing a pair of solid links working through bell cranks and interconnected with the boom and bucket rams. Two circuits only are used for the digging cycle and three for dumping, and increased digging power, without increased energy consumption, is achieved throughout the digging arc. A float position on the boom cylinders means that working equipment takes the path of least resistance reducing bucket wear and undercarriage tipping. Automatic and rapid return to digging position from any height also minimises cycle times.

The RH120C backhoe has a TriMatic system incorporated which, through the two piece boom and related hydraulics, evens out pressure level during lifting and at close reach increases digging and loading height and machine stability.

O & K's smallest machine in its present trio of hydraulic mining excavators, the RH75, was the first to be put into production in 1975; it was developed from the RH60 which was made between 1971 and 1976. By the end of 1981 about 150 RH75s had already been manufactured. The current TriPower 'C' version was brought out in 1983; it is lighter than its predecessor and weighs up to 119 t, has bucket capacities up to 12 m³ ($15\frac{3}{4}$ cu yd) and twin Cummins engines have been replaced by a single one of 447 kW (600 HP). The TriPower system was initially used on the RH40 introduced in 1982 and the first unit of this model then went to the Kingswood coal mine at Cannock, Staffordshire. It was joined in 1983 by the first RH75C replacing an RH75 which had worked 20,000 hours.

Since the latter part of 1979 ten RH75 shovels have been operating at the Ho-Lin-Ho mine in China stripping overburden and loading lignite. The site, in hilly country 1,000 m (3,280 ft) above sea level near the Mongolian border and 1,600 km (994 miles) from the nearest port of Talien, experiences temperatures from $+33°C$ to $-40°C$ – with high winds adding a chill factor. The RH75s incorporate cold weather steel and hot water is automatically circulated through the diesel engines when the machines shut down.

The fourth western European manufacturer to produce a hydraulic mining excavator is Liebherr with its R991. The R991 weighs up to 164 t with bucket capacities up to 12.3 m³ (16 cu yd).

Either two Cummins diesel engines totalling 536 kW (719 HP) or an Alsthom three phase, squirrel cage, AC motor of 570 kW (765 HP), can be used. From the start of series production of R991s in mid-1977 up to the end of 1982 forty-three machines were shipped; five for mounting on barges to do dredging work. The thirty-nine crawler mounted units – including the prototype – were destined for the USA (twelve machines); Australia, West Germany and France (seven each); and one each to Morocco, Greece, Canada, Mexico, Guinea and the UK. R991s have been put to work in France at coal mines and on construction of the new SNCF Paris-Lyon high speed train route, in West Germany on making the River Saar navigable for barge traffic, and on Groote Eylandt island in the Gulf of Carpentaria off northern Australia loading manganese ore onto trucks.

In Romania Promex manufacture the 144 t SC7001 with twin diesel engines producing a total of 567 kW (760 HP). It is available with capacities of 5 m³ ($6\frac{1}{2}$ cu yd) as backhoe or 7.5–10 m³ ($9\frac{3}{4}$–13 cu yd) as shovel.

In the mid-1970s the USSR considered the possibility of creating a range of four hydraulic excavators of 8, 12, 20 and 40 m³ ($10\frac{1}{2}$, $15\frac{3}{4}$, 26 and 52 cu yd) capacities designated 'EG' as shovels or 'EGO' as backhoes. At the same time Poclain 1000CKs were on test in the Uralmashavod Heavy Engineering Works (UZTM). Testing of the first machine to be made, the prototype EG-12, began in 1978 at the Kedrovoye coal mine in the Kuzbas area. This excavator weighs some 260–280 t, has bucket capacities of 10–15 m³ ($13–19\frac{1}{2}$ cu yd), is electrically powered and uses the undercarriage of the EKG-5 cable shovel.

The parameters of the EG-20 were developed at the A. A. Skochinskiy Institute and turned into working specifications by UZTM's planning department. This was the second model in the proposed series to be built; it has bucket capacities of 20–25 m³ (26–33 cu yd), weighs 460–500 t, as a shovel has a maximum digging height of some 18 m (59 ft) and is also electrically powered.

In 1982, it was reported that both EG-12 and EG-20 prototypes were undergoing pre-series production trials. There is, however, no indication that either the 160–180 t EG-8 or the 800–900 t EG-40 has been fully designed or a prototype manufactured.

The largest hydraulic excavator made in the USA is the Marion 3560 weighing up to 272 t and therefore as it matches the Demag H241 in weight, being equal second on the world size scale – excluding Soviet machines. Unlike most manufacturers which produce hydraulic excavators of small and medium size before making mining-size machines, Marion's large 3560 is the company's first fully hydraulic excavator and it has been indicated that smaller models might follow. After preliminary testing at two nearby coal mines the prototype 3560 joined four Marion 201-M cable shovels in 1981 at the Pelver mine, near Paintsville, Kentucky, for final tests before the model was launched in 1982. By the end of 1983 as well as making other pre-series production machines, Marion had shipped three 3560s including two electric backhoes to the Bullmoose coal mine at Tumbler Ridge, British Columbia. The shovel version of the 3560 has capacities of 15.3–22.2 m³ (20–29 cu yd) and the backhoe 12.2–17.6 m³ (16–23 cu yd). Power is produced by two water cooled V-12 Caterpillar diesel engines giving a total of 1044 kW (1400 HP). Unlike many hydraulic mining excavators the engines and hydraulic and electrical components are enclosed in a walk-through machine house.

The operator's cab of the 3560 is fully insulated with thermostatically controlled heating and air conditioning. Radio and tape deck are included as standard fittings, and an annunciator panel alerts the driver to potential problems through an electrical sensing system which monitors twenty-three machine functions. The 3560 has parallel crowd capability and despite the machine's size its modular construction allows assembly in five days.

Koehring had led US hydraulic excavator manufacturers into the 100 t plus class in November 1973 when the first 118 t 1266D backhoe was shipped to St Cloud, Minnesota; over a hundred of these were eventually produced. This model has now been replaced by the 1466FS shovel at 140 t with bucket capacities up to 11.5 m³ (15 cu yd) and the 1466 backhoe – which can take shovel equipment – at 132 t with capac-

ities up to 12.2 m³ (16 cu yd). Both machines have either two turbo-charged twelve cylinder GM Detroit 12V-71T water cooled, two stroke, diesel engines developing a total of 670 kW (898 HP) or two turbo-charged six cylinder Cummins KT1150C-450 water cooled, four stroke, diesel engines developing a total of 619 kW (830 HP); the 1466 can also be fitted with a naturally aspirated variation of the GM Detroit engine. Many features are common to both models – a simplified horsepower controller system and new Electronic Sensing Protection System (ESP) – but only the 1466FS has an elevated cab. At the Bridger mine in Point of Rocks, Wyoming, two 1466s joined a Bucyrus-Erie 195-B shovel and wheel loaders during 1980–81 loading coal into bottom dump haul units.

Northwest's 100-DH backhoe, used particularly for pipeline work, at 107 t just comes within this group of excavators and can be fitted with a coaling bucket of 11.5 m³ (15 cu yd). It was introduced in 1977 and is powered by a V-16 diesel engine.

P & H produce one model of hydraulic mining excavator, the 1200; although a larger 2200 weighing up to 502 t and a smaller 700 up to 94 t, just below the group weight limit, have been promoted but none yet manufactured. The first six 1200s were made in 1979, two more followed in 1980 and a further six in 1981. Although the manufacturer is based in the USA this machine was developed and initially produced at the company's West German factory. The 1200 weighs up to 183 t and has shovel and backhoe, bucket capacities up to 13.8 m³ (18 cu yd). The P & H 'Maxi-Matic' hydraulic system is fitted which uses variable volume, axial piston pumps which are self-regulating. Five of these initial fourteen 1200s are operated by Simms, Sons and Cooke at coal mines in Greater Manchester and Cumbria; all are shovels of 9.2–10.5 m³ (12–13¾ cu yd) capacity – the first went in 1980 to the Amberswood mine at Hindley, the next two to Millers Lane, Atherton, and the final two to Pica.

Although an early entrant into the field of hydraulic excavators, it was 1981 before Bucyrus-Erie introduced a hydraulic mining excavator, the 500-H backhoe. This 102 t machine, of 469 kW (630 HP), has a maximum bucket capacity of 9.2 m³ (12 cu yd) and high

B-E 500-H loads coal at the Seneca Mine, Colorado.

stress 50N steel is used in front end components. A 500-H operates on coal loading duties at the Seneca mine near Oak Creek, Colorado.

In 1982 B-E introduced the 550-HS shovel at 127 t, powered by the same engine as the 500-H, with normal bucket capacities up to 7.7 m³ (10 cu yd). Four-bar front end linkage gives parallelogram geometry for level crowd. Both the 500-H and the 550-H incorporate rust-free Tri-Armor fibreglass for the machinery house. One of the first 550-HS machines was put to work in a basalt quarry in Bound Brook, New Jersey.

Japan's first hydraulic mining excavator was made by Hitachi: the UH801, weighing up to 159 t, appeared in 1979. It has bucket capacities of 8.4 and 12 m³ (11 and $15\frac{3}{4}$ cu yd) as a shovel, and 7.0–10.6 m³ ($10\frac{1}{4}$–$15\frac{1}{2}$ cu yd) as a backhoe. Two Cummins diesel engines operate ten variable displacement axial piston pumps set at 250 bar (3,600 psi) and two gear pumps, which with four control valves enable both independent and combined operation of all functions. Hitachi was one of the first manufacturers to develop an automatic level crowd mechanism for hydraulic

shovels, this is fitted to the UH801 and is operated by a single control lever. Solar Sources operate two UH801 shovels in Indiana and West Coal operate another in Tennessee.

Komatsu followed Hitachi's UH801 with its 160 t PC1500-1 shovel introduced in 1981; by the end of 1983 seven machines had been manufactured. Todaka Mining Industries co-operated on the development of the prototype PC1500-1; the company then purchased the first production unit with an 8.5 m³ (11 cu yd) bullclam shovel bucket for its limestone quarry at Tsukumi, Kyushu, where it works with a Komatsu-Bucyrus 20HT, three Komatsu PC200s and a Mitsubishi excavator. The PC1500-1 at Tsukumi moves some 12,000 t of limestone per day. The PC1500-1 has bucket capacities up to 14 m³ ($18\frac{1}{4}$ cu yd), has either two Cummins or two Komatsu diesel engines, and a 'Computer Aided Optimum Saving Energy' (CAOSE) hydraulic system controlled by an on-board microcomputer.

In about 1982 Mitsubishi introduced the

Hitachi was the first Japanese manufacturer to make large hydraulic excavators; the UH801 appeared in 1979.

MS1600: so far only one of these 165 t machines has been built and is also operating in Kyushu. Backhoe and shovel versions are available, both with maximum capacities of 12 m³ (15¾ cu yd). Two Caterpillar diesel engines are fitted as standard, but a Mitsubishi-Cummins diesel is offered as an option. Again a microprocessor is fitted, and this makes possible intricate digging and loading operations by using a single lever at any one time.

Additional manufacturers will, no doubt, enter this field – in 1984 Kobelco introduced a 130 t shovel, the K975 – and developments of larger and more sophisticated hydraulic mining excavators will continue. There are indications that both Hitachi and Komatsu propose fairly soon to introduce machines larger than the UH801 and PC1500-1 respectively; and a project for a 5–600 t hydraulic excavator, jointly sponsored in Japan, is also believed to be under consideration.

Mitsubishi has produced this 165 t MS1600 prototype, equipped as shovel or backhoe, and on test on Kyushu Island, Japan.

CHAPTER 8
Special Types and Equipment

Despite major changes in the construction, size and power sources of cable excavators since the first practical shovels appeared early in the nineteenth century, the fundamental geometry of a shovel's digging equipment has changed surprisingly little; William Otis clearly got it more or less right from the start. It was not until the late 1960s and early 1970s when George Baron, of Marion, developed the 'Superfront' concept that any really significant improvement took place. The evolution of this concept provides a good indication of how design and technological progress often takes place in the excavator manufacturing industry: initially as an applied research response to users' needs from which further, perhaps unforeseen, product development and application arises.

In 1966 the Peabody Coal Company was interested in the possibility of acquiring an excavator which could dig and load thin seams of coal in Missouri, Arkansas and Oklahoma; without the need for prior ripping or blasting and without leaving coal or loading underlying fire clay. Wheel front-end loaders, skimmers and conventional shovels all had limitations; the ideal solution was a shovel with variable pitch whereby the bucket could remain flat on grade through a long clean-up stroke.

Some Marion 4161s first introduced in 1935, and the 5561 and subsequent Marion models of stripping shovel, had knee-action crowd which gave them longer clean-up radii and lighter and more efficient front-ends than conventional shovels. The knee-action excavator was the obvious starting point from which a design might be created to meet Peabody's requirements. By 1967 design work had progressed to such an extent that a scale model was made to demonstrate kinematics including a pantograph to restrain pitch and a triangular hoist frame and link to replace a fixed boom. Not only was a natural flat

clean-up stroke possible but also this terminated in the bucket rising almost vertically.

A prototype 'Thin seam loader' based on a 101-M electric shovel was shipped to Peabody's Chelsea mine, Oklahoma in 1968. It worked there for several months before going to another mine on the Kansas-Missouri border, from whence it came in 1972 to its present location, a limestone quarry near the Marion factory.

As early as 1968 a scale model of a 15.3 m³ (20 cu yd) general purpose machine, incorporating the new kinematics and based on the 191-M basic design, was constructed. After further tests in 1972 on the 101-M prototype, the following year an existing 191-M was converted into a 194-M by the addition of the newly devised front-end and other modifications. This operated on the Minnesota taconite field for two years before being changed back to a 191-M. A second

Special equipment devised by cartoonist W. Heath Robinson.

George Baron, who devised the Marion Superfront, carries on a tradition of innovative engineering in excavator design dating back to William Otis in the 1830s.

A Marion 204-M Superfront operating at the Eagle Butte coal mine, Wyoming, in 1983.

similarly converted 191-M worked in a copper mine near Tucson, Arizona, during 1975–76. Both 194-Ms provided test-beds for designing a production machine, the 708 t 204-M Superfront, which was based on the 201-M.

The model which had evolved from the 'Thin seam loader' is the only cable electric mining shovel with a variable pitch bucket. Pitch control is achieved by two overlapping parallelograms with a common bellcrank which transmits any change made to one to the other. Increased digging power results from the triangular hoist frame acting as a lever and applying force to the bucket, and hoist and crowd motions – which often oppose each other on conventional shovels – work together through the frame; a relatively large bucket can be fitted as the front-end is lighter than normal; better bucket filling and dumping is possible; and the original design intention of a long level clean-up pass is achieved.

Between 1976 and 1978 five 204-Ms were built under licence by Sumitomo in Japan, followed by five more in 1981; all with 19.9 m³ (26 cu yd)

buckets and destined for work in the USSR. These were all shipped to the new Neryugrinskoye coal mine on the south Yakutsk coalfield in eastern Siberia, where a 300 Mt deposit of high grade coking coal is being exploited jointly by the USSR and Japan. Six 18.3 m³ (24 cu yd) 201-Ms have since joined the ten 204-Ms at the mine where production is planned to reach 13 Mt annually. In 1979 two US-built 204-Ms started work with 22.9 m³ (30 cu yd) buckets and cable operated crowd (unlike the 194-M prototypes and the machines built in Japan which had hydraulic cylinders operating the crowd). One is at Amax's Eagle Butte coal mine, near Gillette, Wyoming, the other at Energy Fuel's mine at Energy, Colorado.

In the late 1960s while George Baron was devising the Superfront in Marion, Ohio, on the other side of the Atlantic Norman Brocklebank was also at work applying new ideas to front-end geometry, in Priestman's factory at Hull, Humberside. This innovative work, though, to reproduce the crowd action of a cable machine on a 1.7 m³ (2¼ cu yd) hydraulic shovel, did not meet with comparable success. Although a single 'Bison' was produced in 1967 no more units were made. However, fifteen years later Norman Brocklebank's determination to devise a machine with improved front-end kinematics paid off: in 1982 the first model of the Priestman VC range of

machines was put into series production. The breakthrough had been not with a shovel design, but with a machine incorporating both cable and hydraulic components, and combining elements of the dragline and backhoe type of equipment which had remained fundamentally the same since they were first conceived at the end of the nineteenth century.

To achieve balance on an excavator the sum of forward moments by its front-end equipment with a full bucket must equal the rearward moments resulting from the weight of the base machine and its counterweight. Although a machine's capacity and/or boom length can be increased with disproportionately smaller increases in total weight, nevertheless for much of the operating cycle excavators are utilizing unnecessarily large counterweights with a centre of gravity excessively far back. Therefore when slewing, power is being wasted for much of the time, and a fluctuating centre of gravity causes instability. Consideration of this situation gave rise to an initial concept, the variable counterweight (VC) working automatically in sympathy with the front-end equipment. Both B-E and

The unique Priestman VC15 variable counterbalance machine incorporates both dragline and hydraulic excavator features; in Pakistan this unit clears a drainage canal.

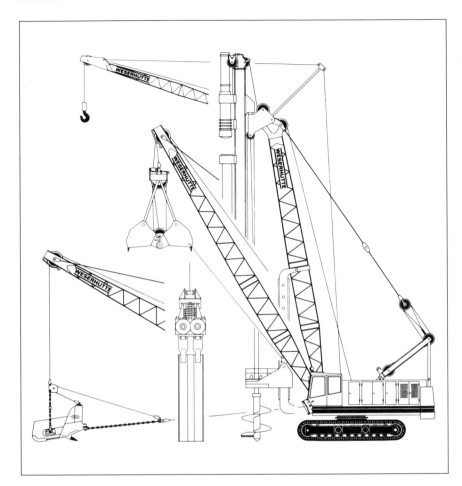

Some of the many types of equipment that can be fitted to a lattice boom extractor; grab; crane a Weserhütte base machine: dragline; pile extractor, grab; crane with fly jib; pile driver; and earth auger.

Marion produced stripping shovels with counterbalanced hoists, but these merely assisted the electric motors until motors with improved torque were devised, and did not alter the machine's centre of gravity.

Advantages and disadvantages of the front-end equipment of draglines and hydraulic backhoes were compared and the advantages of each incorporated into one design: the good outreach, low front-end weight and direct pull to the bucket of a dragline with the ease of operation, safety, controlled movement and arm-assisted penetration of the hydraulic backhoe.

A combination of the VC concept and the dragline/hydraulic backhoe design resulted in the VC15 which has a light box-boom incorporating a tail on which the counterweight slides to alter the centre of gravity. This counterweight, activated by a hydraulic ram, in turn takes an equally light arm out by means of a cable and

quadrant giving constant torque. Wrist action on the bucket is achieved by means of a hydraulic ram, and a hydraulic winch and cable gives direct pull to the bucket for digging. The two basic motions, pulling the bucket in and taking the arm out, are interconnected and controlled through one lever by the driver.

The VC15 weighs up to 20 t and at its maximum outreach of 15 m (49 ft 3 in) the rated load is some 50 per cent greater, and the machine about 15 per cent lighter, than a comparable conventional excavator. Buckets range from 0.6 to 1.1 m³ ($\frac{3}{4}$–$1\frac{3}{8}$ cu yd) capacity and the maximum digging depth, with an optional cranked boom, can be as much as 9 m (29 ft 6 in). A low ground pressure 'Bogmaster' version of the VC15 appeared together with a 26 t VC20 in 1984. Proposals have also been made for a 9 t VC8, 13 t VC12 and 36 t VC25 with a lattice boom. A prototype hydraulic backhoe only incorporating the VC concept has been made for development testing.

By the end of 1983 over thirty VC15s had been produced and were operating in places as far apart as the UK, Mozambique, Iran, the Netherlands, Spain, Thailand, Australia and the USA. VC machines, originally devised for long-reach land drainage and irrigation work, have been found to be applicable to a wide range of general excavation tasks including clearing power station flyash lagoons, sand and gravel extraction, land reclamation and underwater trench excavation.

With the substitution of hydraulic machines for many small and medium size cable excavators in recent years, it is not surprising that most current or recent special types of excavator are hydraulic ones. The original telescoping boom excavator was probably the Gradall devised by Ray Ferwerda, a contractor and part-time inventor based in Cleveland, Ohio. He built only four machines; the initial model consisted of a telescoping boom with a blade to help grade slopes. In 1945 the Gradall manufacturing rights were purchased by Warner & Swasey and developed with the addition of advanced hydraulics and a variety of buckets; by 1975 over 8,000 Gradall's had been made. Now, including W & S, this type of specialised hydraulic excavator is made by at least seven manufacturers and former makers include Link-Belt, Marion-Quickway and Schwing in the 1960s. Telescoping boom excavators perform five basic movements similar to the human arm and wrist: boom telescoping to reach in and out, boom raise and lower with the ability to dig straight down, full circle slew, boom tilt and bucket wrist-action pivoting.

There are currently four basic W & S Gradall models, the smallest the G-3 is in R-wheeled rough terrain, XR- with railway guide wheels, and W- truck mounted, versions; the G-660C, G-880C and the largest, the G-1000, are truck or crawler mounted. G-1000s have digging depths of 5.6 m (18 ft 5 in) and surface reaches of 10.5 m (34 ft 6 in). Remaining US manufacturers are Badger with 300 and 460 Hydro-Scopic models either truck or crawler mounted; Koehring has Bantam Teleskoop models, the truck mounted T-644 and T-747 and crawler C-744; and Little Giant just has the 34TX truck or crawler mounted machine.

EWK's Combicraft PL820R is a crawler mounted West German machine also made with a jack-up undercarriage for tunnel work, and in

Telescoping boom excavators, such as this W & S Gradall G660, are well suited to road works.

Swiss-made Menzie Muck machine exhibits its abilities as a climbing type of excavator.

Probably the smallest excavator ever made is the 0.8 t Powerfab 360W which is manufactured in the UK.

MF form is mounted on a Tatra truck chassis. In Italy Benati make the crawler IM14 and in Czechoslovakia Detvan's range includes the UDS114-a (introduced in 1983) and UDS113-a (introduced in 1977) both on Tatra chassis, together with the crawler UDS110-p (introduced in 1977).

Small hydraulic climbing backhoes are characterised by their possessing two unpowered wheels and two hydraulically activated legs; the sophistication of these machines, and their range of movements of wheels and legs in conjunction with the digging equipment to provide travel, varies according to manufacturer. Swiss-made Menzie Mucks, over 4,000 of which have been produced since 1966, were introduced primarily for installing towers for powerlines in mountainous areas. The machine is capable of manoeuvring on 65° slopes and levelling itself on 45° slopes – as well as working in 1.8 m (6 ft) of mud and water. In 1979–80 a Menzie Muck put in the foundations for the 105 towers of the first six chair lifts at Beaver Creek ski area, Colorado; one has also worked at Macon in swamplands alongside the Ocmulgee River, middle Georgia. An

optional powered third wheel unit can be added to make a 5000 version which can propel itself over long distances. The climbing Tigrotto, Tigrone and Buggy models are made by Euromach, and the Kamo 3X by Moro, all in Italy; Schaeff's HS40A comes from West Germany; and in the UK Smalley manufactures the 5 series 3 and Powerfab the 360W. The 0.8t Powerfab machine is probably the smallest full slew single bucket excavator in the world. It has normally a 7.4 kW (10 HP) Honda petrol engine or a 6.3 kW (8.4 HP) Petters diesel engine and its maximum digging depth is 2 m (6 ft 7 in). Some 2,100 of the 0.08 m³ ($\frac{1}{10}$ cu yd) buckets of the 360W would be needed to fill the 168 m³ (220 cu yd) bucket of the world's largest excavator, 'Big Muskie'. Unlike 'Big Muskie' which can fit a complete house in its bucket, the 360W can easily travel through a house doorway!

Brøyt excavators are unique amongst the current ranges of conventional hydraulic machines because their undercarriages are not driven. Each unit is mounted on two rubber tyred wheels and two steel drums with studs; travel is achieved by using the front-end equipment. In addition the

Dropped a bit! Brøyt excavators have undriven undercarriages and are particularly suited to quarry and tunnelling work.

two piece boom is mounted so that its inner section is pivoted behind the slew ring with the lifting ram in front. Brøyt's current range consists of the 21 Series II backhoe weighing up to 13 t, the X41 shovel up to 33 t, and the X50 shovel up to 47 t and equipped with a 4 m³ ($5\frac{1}{4}$ cu yd) bucket. Both the X41 and X50 have electric versions used particularly for tunnelling operations.

Brøyt excavators have worked on Norwegian hydro-electric schemes in Sira-Kvina, Aurland and Eidfjord as well as on similar schemes in other parts of Europe and in Africa. The St Gotthard and Seelisberg Tunnels in Switzerland, the Fréjus Tunnel between France and Italy, and the Katschberg and Arlberg Tunnels in Austria have all had this special type of excavator involved in their construction or associated works. In Norway one machine loads pyrite 450 m (1,476 ft) underground, another loads olivine at a surface mine, and a dismantled unit was airlifted by helicopter 7 km ($4\frac{1}{3}$ miles) for road building on the mountains. A Brøyt excavator digs limestone in Alicante, Spain, and yet another works in a granite quarry on the Danish island of Bornholm.

The Pingon excavator ('Sitting Bull' in the US market) is also distinguished by its form of mounting. The chassis consists of a main fabricated X-form frame on which is mounted the slew ring and conventional upperworks. From the sides of the frame extend two leading and two trailing cast arms carrying large drive wheels. Each inner pivoted end of the arms consists of a quadrant type rack which engages a pinion on the chassis. To travel the pinion turns to lower the arms, thereby lifting the chassis and placing the wheels on the ground – steering is by the skid steer method. When the machine is in position to work the arms and wheels are raised to lower the chassis and upperworks, and the excavator operates resting solely on its X-form frame. One model, the 14 t 14C is made with CH-deluxe cab, and TR – adjustable boom versions, and can be equipped as a 0.2–0.99 m³ ($\frac{1}{4}$–$1\frac{1}{4}$ cu yd) backhoe or 0.78–1 m³ (1–$1\frac{1}{4}$ cu yd) shovel.

Early Schwing Exakt 431 backhoes had a straight line grading system achieved by means of a complex front-end design; subsequently the 443 – now developed into the 15 t wheeled M450 and 22 t crawler R444 – initiated use of a more conventional boom with parallelogram kinematics giving parallel movement at the bucket to plus or minus 2 cm ($\frac{3}{4}$ in). Bucket tilt is standard on the backhoes and both models can be used as shovels.

Schaeff excavators have 'Knickmatic' devices attached to the bases of their booms which give stepless lateral adjustment over the entire width of the machine within the rear swing circle. This results in the machines being able to achieve full digging depth parallel to, and alongside, their crawlers.

Poland's Brawal 1611 crawler backhoe, named after those who devised it – Professor Ignacy Brach and Ryszard Walczewski – was patented in 1977. A 30 t prototype with a 1.6 m³ ($2\frac{1}{8}$ cu yd) bucket was produced by Warynski in 1980 based on a Menck-Koehring M250H undercarriage. The patent mainly covers the fitting of a normal boom which, at its base, can be moved horizon-

ally to correspond with the motion of a telescoping boom excavator; also a new type of shock absorber is fitted which assists hard digging. The machine's operational features include the ability to make trenches with level bottoms and decreased energy consumption because of the new kinematics and the use of an energy recovery and stabilisation system.

In past years a number of novel types of hydraulic excavator were in existence for a time, with varying degrees of success. In 1958 Koehring introduced the cable/hydraulic 205 Skooper crawler loading shovel with level crowd from which was developed the fully hydraulic 505 and

Schaeff excavators are fitted with 'Knickmatic' devices allowing digging parallel to, and alongside, their crawlers.

666 Skoopers. Blaw Knox in 1959 brought out a similar Hydroscoop – but it was possible to make both cable and hydraulic elements work against each other and bend the boom! In 1969 B-E made a prototype 3.8 m³ (5 cu yd) 60-S hydraulic loading shovel mounted on tank-type crawlers capable of speeds up to 10.8 kph (6¾ mph). Between 1978 and 1980 Northwest manufactured the 65-DHS hydraulic shovel with its unique 'Dig Link' geometry to increase crowd and breakout forces and a 'Transtick' hydraulically extendable and retractable bucket arm.

Although designed primarily as earthmoving tools the smaller sizes of excavator are incredibly versatile base machines capable of accepting a very wide range of equipment for additional applications. Even when used for their prime purpose, there is available a vast array of special buckets and other excavating equipment in addition to the standard shovel, backhoe and dragline front-ends. Trenches and ditches, for example, are often dug with V- profile or trapezoidal buckets.

Various types of mountings provide additional versatility to conventional excavator upperworks. Barge-mounted backhoes or grabs are often used for dredging. A Manitowoc 3900 with a 1.9 m³ (2½ cu yd) clamshell grab operated off Miami Beach, Florida, on construction of the Virginia Key outfall sewer extension; and another Manitowoc, a 4600 Series I with a 5.7 m³ (7½ cu yd) clamshell, dredges the Allegheny River in Pennsylvania for aggregate. A Liebherr 991 backhoe has worked off the Mediterranean island of Sicily excavating a trench for a section of a pipeline from Tunisia to Italy; in 1977–81 a Poclain 1000 backhoe created a new access channel in the Baltic Sea to the Finnish port of Rauma; and Demag H241 backhoes work as dredgers in Austria and the USSR.

Hitachi and Mitsubishi have produced amphibious soft-terrain excavators mounted on exceptionally wide crawlers. An Hitachi MA125U, equipped as a dragline, was used on canal widening in Iraq at temperatures of 50°C; another of the company's specialised machines, the UA04, can work automatically under water with remote control. Richier's marsh special H43 uses wide polyurethane track pads on an H45 undercarriage to obtain its very low ground bearing

pressure. Such machines can often travel and work on ground that would not support the weight of an average person.

Rail-mounted excavators include the Benati 14 t 125 and 16 t 145 Railroad models which have additional oscillating axles with bogie wheels. When the machine is used on a railway these are lowered by hydraulic rams to lift the road wheels, which in turn provide drive by coming into contact with the bogie wheels, locked into position. Grapples for lifting sleepers and clamshells for ballast handling are amongst the commonly used railway equipment. Atlas specialises in rail-mounted excavators, such as a 1302 DK used for maintenance work on the Berlin underground system.

Excavators can be operated without drivers on board. Hitachi have built a UH04R shovel to perform dangerous jobs, such as handling hot slag, by radio control. Åkerman adapted an H9MB for remote control when a machine, equipped with a hammer, was required to work near the top edge of a quarry face: an electric cable connected the machine to a portable control box operated by the driver standing in a safe location.

Most small and medium sized draglines can also be used as cranes or grabs including specially mounted and adapted types, for example for dockside use. Even hydraulic backhoes are sometimes equipped with booms suitable for such work: the Hymac 1080 could lift 11 t with a lattice boom and the Mitsubishi MS280 has been fitted with a box-section boom for crane work. R-B, which has developed base machines into a series of 'Supercranes', in the late 1970s shipped fourteen 30-RBs and four 61-RBs equipped as cranes for the Sidor steel mill expansion project in Venezuela. The FMC Link-Belt LS-518 is also a 136 t capacity crane; the NCK-Rapier 1405D a 91 t crane, and the American 4120 a 25 t crane.

Pile-driving and boring with earth augers for foundation work are often carried out by excavators equipped with the relevant machinery. Eleven NCK-Rapier crawler machines were used in 1979–81 to place 1140 reinforced concrete piles for a new bridge over the River Orwell in Suffolk; in the same contractor's fleet is an Ajax C75 fitted with a Hughes 100KCA rig

EXCAVATORS

Pontoon mounted 54-RB, used as a crane or grab on an outfall sewer scheme, shared Almiros bay with holidaymakers on the Greek island of Crete in 1982.

Åkerman H-W H9MB re-handling wheel sets for Atchison, Topeka & Santa Fé Railroad at the Topeka workshops, Kansas.

capable of producing a bore up to 3 m (9 ft 10 in) in diameter and 55 m (180 ft) deep. On the new SNCF Paris-Lyon high speed railway in France several thousand foundations for the catenary supports were made by two wagon-mounted Poclain 75s modified to lessen the swing radius and give clearance in tunnels.

To undertake demolition work by excavators, equipment ranges from the use of pneumatic or hydraulic hammers fitted to otherwise standard machines and steel drop balls fitted to cable machines with lattice booms, to the use of machines specially adapted. Available equipment for hydraulic excavators includes long arms with deflector plates, adjustable demolition grabs, armour-plated cab windows and undercarriage guards. Hymac in conjunction with the UK's Building Research Station developed a hydraulic 'Nibbler' whose pincer movement could bend and snap concrete up to 0.38 m (1 ft 3 in) thick. Hydraulic hammers can, of course, also be used as rock breakers: two Poclain 75CLs fitted with such equipment teamed up with two 75P grabs in Qatar to dig a section of the trench for a water pipeline from the Ras Abu Finta desalination plant to the industrial town of Umm Said.

Steel works have special 'excavation' needs; Benati caters for these with a range of eight specially equipped machines for stripping refractory linings of converters, electric furnaces, ladles and runners. Rotating, articulated and telescoping booms are used with ripper teeth, buckets and hammers. Benati's BEN160CSA was developed for use at the Piombino Steel Mill in Italy, and Poclain was commissioned by Creusot-Loire to produce special equipment incorporating a rotating boom on the 75CL for the company's French steel mills.

Extensive use of excavators equipped with electro-magnets and multi-tine grabs is made by the scrap handling industry. One typical application is that of a Unit H-471 at Berlin, Wisconsin, which can load up to 20 t of scrap metal into a truck, with its 1.7 m (5 ft 6 in) diameter magnet, in about ten minutes. The logging industry also depends heavily on excavator-based machines. Grapples for log loading and handling, delimbers with telescoping booms and brush rakes are amongst Caterpillar products; Case makes the Drott 40LC Feller/Buncher and 45 Cruz-Air

NCK-Rapier 305A, equipped as a crane with a fly jib, moving shuttering on a building site at Martlesham, Suffolk.

logger; and Koehring's line of five logging machines can be fitted with any one of three types of feller/buncher head and have an optional mechanism to elevate the cab while operating.

Sugar beet is harvested in autumn in central Europe by a Liebherr A911B and an A921B, and sugar cane in Hawaii by five Bucyrus-Erie 30-Bs; stagnant water and undergrowth breeding grounds for mosquitos are eliminated high in the French Alps by a Poclain TCB, and a Caterpillar 215 works deep in an underground iron ore mine dislodging rock from the ceiling after blasting.

The military use of excavators covers normal models on standard excavation duties, and specially adapted excavators or excavator-derived machines used for specifically military purposes. Time has not always been on the side of British attempts to make purpose-built military excavators: in 1918 two Priestman-built steam grabs mounted on tank chassis reached the Western Front just as the Armistice was signed; and 'White Rabbit No 6', later codenamed 'Cultivator No 6' but more usually referred to by its builders at Ruston-Bucyrus (whose predecessor

The NASA crawler transporter at Cape Kennedy, Florida, was built by Marion using stripping shovel technology.

company had built the first tank in 1915) as 'Nellie', never saw service in World War 2. Nellie, first conceived by Winston Churchill in World War 1, was designed to dig trenches for undercover movement of personnel and vehicles between opposing defended lines. Production had just started in 1940 of two models of this crawler machine – weighing up to about 130 t and of 894 kW (1200 HP) – when German forces circumnavigated the Allied Forces' Maginot Line and made the intended breeching of the German Siegfried Line by Nellie irrelevant.

The 'SP Armoured Engineer Vehicle 2' (GPM) is currently in use with the West German Federal Defence Forces – the Bundeswehr – particularly for bank and in-water preparation work to enable deeply wading vehicles to cross water areas. Elements of two EWK telescoping boom excavators, each with 180° slew, are mounted on the front of a Leopard 2 tank chassis fitted with a dozer blade. Following the Falkland Islands conflict of 1982 the British military authorities purchased for use on the islands six Hymac excavators – two 590CTs were modified for land mine clearance with wide and hollow track pads, blast-protected cabs and rotavator attachments, and another was fitted with remote control – and ten Manidig mini backhoes.

Military activities indirectly gave rise to another special machine when, as a result of equipment shortages following World War 2, Chuck Doerr designed and built at a coal mine at Taylorton, Saskatchewan, during 1947–48, a single 5.3 m³ (7 cu yd) 7-D electric shovel. This was constructed from B-E 320-B stripping shovel and Bucyrus-Monighan 5-W walking dragline parts, and was probably the first shovel made with lower frame independent electric propel.

When natural or man-made disasters occur excavators will often be found doing invaluable emergency work to save life or repair damage. In 1966 at the scene of one of the UK's worst disasters when colliery spoil slid down a valley side burying houses and a school at Aberfan, South Wales, Ruston-Bucyrus machines worked around the clock; after the 1980 eruptions of Mount St Helens in the state of Washington five B-E 88-B and five Manitowoc 4600 draglines were amongst the excavators that cleared debris out of the Cowlitz River and North Toutle Dam; and in 1983 a Poclain backhoe was amongst the rescuers immediately after the demolition by a bomb of a building occupied by troops of the French peace-keeping force in Beirut, Lebanon.

Two manufacturers have recently promoted futuristic designs of excavator. The concept of Hitachi's UH999 was demonstrated in 1981 when the company stated a prototype had been built. This large hydraulic mining machine would have a rotating bucket and telescoping dipper arm allowing dual use as a shovel or backhoe without changing equipment; a cab which could move hydraulically to improve the driver's view; and crawlers which could be extended for increased stability. O & K's 'Futura' backhoe prototype was demonstrated in

1983; features include fuel conservation by automatic reduction of engine speed after a period of time when no power is being delivered and by regaining the energy resulting from braking after slewing; retractable side supports to increase stability and lower the ground bearing pressure; and operation by movements made by the driver of a single 'Syncro-pilot' unit, which conforms to those of the standard backhoe, and which incorporates a computer display in full text of a system for acquiring and evaluating operational data.

A framed copy of a telegram in Marion's offices gives a clue to the most unusual 'excavator' of all: this telegram was sent to the first men to journey to the moon and says 'Have the first three miles on us'. In the mid-1960s Marion's expertise in designing and building lower works for large stripping shovels was utilised to provide the two giant crawler transporters used by the National Aeronautics and Space Administration (NASA) at Cape Kennedy, Florida. The transporters were delivered in 1966 and since tests the following year these carried from the assembly building to the launching site – perfectly balanced – all the complete Saturn V-Apollo rockets and spacecraft with their portable launching pads and 136 m (445 ft) high towers for moon shots and space flights; and more recent rocket and spacecraft assemblies including the Space Shuttle. Each transporter is 39.9 m (131 ft) long, 34.7 m (114 ft) wide and 6.1–7.9 m (20–26 ft) high; weighs some 2,700 t unladen and carries at least an equal weight when loaded; has six diesel generator sets providing electric and hydraulic power; and its four pairs of 2.1 m (7 ft) wide crawlers can move it at up to 1.8 kph ($1\frac{1}{10}$ mph).

These excavator-derived units which helped to put man into space in turn repaid the industry which gave birth to them. Marion engineers had designed gearing with an internal, involute spline of an unprecedented size for the transporters and to produce them it was necessary to develop a special manufacturing tool. These improved gears were later introduced into Marion mining shovels and draglines.

CHAPTER 9

Manufacturers and their Products

The modern excavator manufacturing industry is a highly complex operation involving at least 150 factories in some thirty countries. Very early characteristics have been retained including manufacture under licence in countries outside the home-base of the original designer-manufacturer; machines entering world trade in significant numbers; the making of excavators often being but one element of a manufacturer's product line; regular changes in makes available as companies close, new ones emerge and take-overs and amalgamations take place; and specific expertise brought into a company by employing former employees of other excavator manufacturers. As would be expected additional features, common in present-day industries of this type, are exhibited; many manufacturers are often part of national groups or multi-national concerns with very diverse product lines; marketing and overall design have major roles alongside technical development and manufacturing operations; and sophisticated techniques – including computer assisted design and manufacturing, and increasing application of modern methods of series production involving the use of robots – are to be found.

Excavator manufacturing does, though, have its own peculiar features. Predominant amongst these is the fact that the industry is concentrated primarily within the USA, Europe and Japan; and the manufacture of certain types of excavator is even more concentrated. The bulk of the world market for large walking draglines and large electric mining shovels for surface mining oper-

The growth of Ruston-Bucyrus and Bucyrus-Erie.

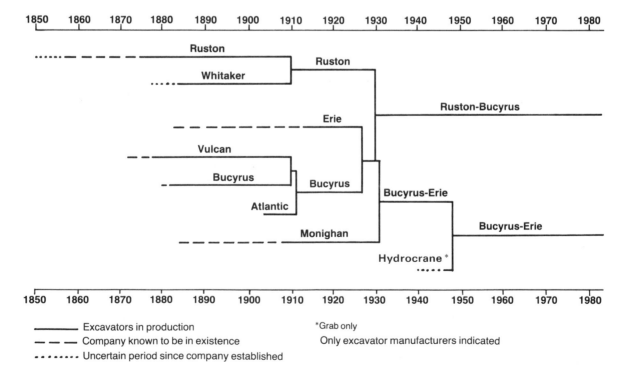

Excavators in production
Company known to be in existence
Uncertain period since company established

*Grab only
Only excavator manufacturers indicated

ations is supplied by only four manufacturers – Bucyrus-Erie, Marion, Page and P & H. Not only this, but these are located in a single region of one country, in Wisconsin, Illinois and Ohio in the USA, and no more than 483 km (300 miles) apart. Another feature is the increasing use of 'badge' engineering, a system whereby a company markets a range of excavators produced by another manufacturer in modified, or essentially original, form. The livery and decals are changed and the machines are sold as products of the licensee, as an additional line if the licensee is not already an excavator manufacturer, or as supplements to its existing range if it is.

ALGERIA
SONACOME   Algerian Government
In the early 1980s a factory was due to open to make hydraulic excavators and cranes.

ARGENTINA
TORTONE   Tortone SA
Since 1969 Benati excavators have been made under licence. Currently nine models of wheeled and crawler machines are made at Córdoba. Tortone also manufactures other items of construction equipment and tractors.

AUSTRIA
LIEBHERR   Liebherr-Werk Nenzing Ges mbH
Liebherr-International AG (Switzerland)
At the Nenzing factory in Voralberg, opened in 1976, Liebherr hydraulically driven draglines have been produced since 1981: currently the HS 840, 850 and 870.
SCHWING EXACT   Friedrich Wilh Schwing GmbH
(West Germany)
The company was founded in 1934 by Friedrich Schwing in Herne, West Germany to manufacture small items of construction equipment; subsequently concrete handling equipment has been a speciality. The first excavator, a telescoping boom TL10, was made about 1961, and the first Exakt type in 1965. The present 15 t M450 and 22 t R444 are made at St Stephan.

BELGIUM
CATERPILLAR   Caterpillar Belgium SA

Caterpillar Tractor Company (USA)
The Gosselies factory north of Charleroi, was opened in 1968 in an area of good communications with a declining coal industry. All four Caterpillar models of excavator are made, together with other items of construction equipment including wheel loaders, dozers and engines.

BRAZIL
DEMAG   Mannesmann Demag Ltda
Mannesmann AG (West Germany)
Demag Equipamentos Industriais Ltda was established in 1974, and the name changed in 1981. In 1977 a new factory was opened at Vespasiano, 20 km ($12\frac{1}{2}$ miles) north of Belo Horizonte in Minas Gerais, originally producing industrial machinery. A prototype H-71 was tested in 1980–81; in addition the H51/H55 and H121 have been made and it is proposed also to make the H241.
FIATALLIS   Fiatallis Latino Americana SA
In Contagem, Minas Gerais, the S90 crawler mounted backhoe, weighing up to 18 t, is made.
FNV BUCYRUS   Fábrica Nacional de Vagões SA
Bucyrus Equipamentos de Construcao Linotatada (BECO), a wholly owned subsidiary of B-E, and FNV, agreed in 1964 to jointly produce B-E equipment; the first excavators were manufactured in 1965. Now FNV, whose products include rolling stock and vehicle components, makes the 22-B at its Cruzeiro factory, São Paulo.
LIEBHERR   Liebherr Brazil Ltda
Liebherr International AG (Switzerland)
The Guarantinguetà factory, between São Paulo and Rio de Janeiro, was established in the early 1970s to produce cranes. It has been indicated that the R991 excavator is to be made in Brazil.
POCLAIN   J. I. Case do Brasil
Tenneco (USA)
Poclain do Brasil was founded in 1972. Manufacturing operations commenced in Conselheiro Lafaiete, Minas Gerais, the following year using imported hydraulic components for LY2P, TCS and LC80 models; in 1974 the TY2P and SC150 were added. In 1977 J. I. Case purchased the Poclain factory and now at Case's Sorocaba factory south of São Paulo, established in 1971, LY2P, LC80 and SC150 models with over 90 per cent Brazilian-made content are manufactured.

VILLARES Equipamentos Villares SA

Since 1965 Villares has produced at São Bernado do Campo, São Paulo, P & H construction-size excavators. Currently crawler and truck mounted draglines are made.

CHINA

Chinese Government
TAIYUAN FACTORY

This is the main production unit in China situated in Shanxi Province in the north-west; the Soviet-designed EKG-4.6 shovel and a small walking dragline, probably also Soviet-designed, are produced here.

FUSHUN FACTORY

Situated near Shenyang (Mukden), in Liaoning Province in the north-east, the 12 m³ (15¾ cu yd) WD1200 shovel is made as well as the EKG-4.6 at the factory.

CHANGJIANG FACTORY

At this factory in Sichuan Province in the south-west the WY160 hydraulic shovel is manufactured.

SHANGHAI CONSTRUCTION
MACHINERY FACTORY

Located in the south-east, hydraulic excavators have been produced here since 1972. In 1979 there was an annual output of some 300 units of a single 1 m³ (1¼ cu yd) model equipped as shovel or backhoe. In 1982 production commenced of the Liebherr R942.

HANGZHOU HEAVY MACHINERY
FACTORY

Also in the south-east, this factory in 1979 made two models of cable shovel; the largest of which had a bucket capacity of 2.5 m³ (3¼ cu yd). A prototype of a 2 m³ (2⅝ cu yd) hydraulic shovel had also been produced at that time, possibly the W2-200.

BEIJING FACTORY

In the capital city small excavators are made; possibly including the WY25D and 0.4 m³ (½ cu yd) WY40 hydraulic backhoes.

Also within China the 10 m³ (13 cu yd) WK-10 is manufactured and, following an agreement reached in 1983, production commenced in 1984 of the Demag H55 and H85 machines.

CZECHOSLOVAKIA

ZTS-DETVA (Detvan) Závody Ťažkého Strojírenství

Czechoslovakian Government

Following the introduction of the Satur 050-K in 1966, the Podpolianské Strojárne works at Detva has subsequently produced other models of telescoping boom excavator. Three current models have bucket capacities of 0.63–1 m³ (⅞–1¼ cu yd); the UDS114-a replaced the UDS110-a in 1983. Products of this factory include wheel loaders, torque converters and gear boxes. The ZTS Group, based at Martin, is the country's major manufacturer of construction machines and has diverse other products.

ZTS DUBNICA Závody Ťažkého Strojírenství

Czechoslovakian Government

The factory at Dubnica nad Váhom was originally established in 1937 as a branch of the Skoda Holding Company of Plzeň. As well as excavators, hydraulic motors and pumps and products for the metallurgical industry are made. From 1978 to 1980 213 Poclain TY55-2, TC45-1 and TY2P machines were manufactured. Since 1980 ZTS Dubnica has developed a basic 0.4 m³ (½ cu yd) backhoe in crawler (DH111), four-wheeled (DH112) and three-wheeled (DH113) versions. By the end of 1982, 1,391 DH113s and a total of 206 DH111s and DH112s had been produced. A larger crawler DH211 and a four-wheel drive DH132 are under development. Hydraulic components are manufactured at Brno and Bratislava, booms are fabricated at Bardejov and Zetor engines come from Zbrojovka Brno.

UNEX Uničovské Strojírny Vítkovice, Železárny a Strojírny Kl Gottwalda

Czechoslovakian Government

Gottwald Vítkovice iron, steel and engineering organisation of Ostrava started production of hydraulic excavators at Uničov in 1970. Previously cable machines were manufactured, the sole remaining model being the 115 t E303. Three models of crawler mounted hydraulic excavator, in shovel or backhoe form, currently in production are the DH103 up to 28 t (which replaced the DH101 and 102), the DH411 up to 28 t and the DH 611 up to 42 t.

EAST GERMANY

BAUKEMA (NOBAS)
VEB Schwermaschinenbau NOBAS
East German Government

Cable excavators, such as the UB 60, 80, 83, 1212 and 1213 universal machines of the 1960s, used to be made at the Nobas factory at Nordhausen. Recent hydraulic models include the crawler UB631 and UB1232, both available as shovel or backhoe, and the hydraulically driven UB1252 dragline version of the UB1232. Graders, rollers and equipment for processing cement and bituminous materials are also marketed under the Baukema name.

TAKRAF (ZEMAG)   VEB Eisengiesserei und Maschinenfabrik Zemag
East German Government
Although established in the latter half of the nineteenth century this factory has only been a manufacturer of cable excavators, at Zeitz, since it produced the 0.9–3 m³ (1⅛–4 cu yd) UB162 probably in the late 1950s. Now the UB1412 universal cable machine is made with capacities up to 2.5 m³ (3¼ cu yd).

FINLAND
LÄNNEN   Lännen Tehtaat Oy
Founded in 1950 as a local community enterprise to build sugar beet and fodder plants at Iso-Vimma, Satakunta, between Turku and Tampere, Lännen has diversified into food canning, plant raising products and engineering. The engineering factory, established in 1959, produces backhoe loaders, forage harvesters and, since the early 1970s, hydraulic excavators. Four models with weights from 14–21 t are produced: the crawler T14 and its M12 wheeled version together with the crawler T18 and its M16 wheeled version.

LOKOMO   Rauma-Repola Oy
This company was created by merger in 1951 and has engineering, shipyard and timber processing divisions. The Lokomo works in Tampere has been producing heavy machinery since 1915: as Lokomo Oy it produced 'Terasmies' universal cable excavators, such as the J series, from the JN2 with capacities up to 0.3 m³ (⅜ cu yd) to the JR92 with capacities up to 3.2 m³ (4⅛ cu yd). In the late 1970s the 20 t crawler T325C and wheeled M325C, and 31 t crawler T340C hydraulic excavators were produced; since then the 21 t T323 and 24 t T326 have been made. In 1980 Lokomo took over the excavator business of ARA, a company which had manufactured hy-

draulic excavators at Turku since the introduction of the 10 t wheeled AK31 in 1965. Cranes, forest machinery, graders and crushing plant are also made. Lännen will make Lokomo excavators from 1985.

FRANCE
LIEBHERR   Liebherr France SA
Liebherr-International AG (Switzerland)
Liebherr's French subsidiary was established in 1961 and the Colmar factory built the following year; the first excavators made were wheeled and crawler mounted machines of the 700 series. At Colmar the following crawler models are produced 912, 922, 932, 942 (and wheeled version), 952, 962, 972, 982 and 991.

PINGON (SITTING BULL)
Ecomat SA
Manubat Pingon established a factory at Belley, Oise, in the foothills of the French Alps in 1957 to manufacture tower cranes. Production commenced in 1958 and in 1969 the unique Pingon excavator based on an X-form skid steer chassis was introduced; this was designed by Manubat Pingon although he had relinquished control of the company six years earlier. From 1977, when an industrial consortium called Cif Loire took over the company, until 1981, only excavators and related cranes were produced; backhoe-loaders and military multi-purpose loaders were then added. In 1979 Pingon became part of the IBH Group which in turn went into liquidation in 1983; Ecomat, established by J-M Obry, took control in 1984. One 14 t model is produced in three versions: the 14 C was introduced in 1971, the 14 CH in 1976 and there is a 14 TR which is a 14 CH with adjustable boom. A 14 D rough terrain mobile crane utilises the basic excavator.

POCLAIN   Poclain SA
Major shareholder, Tenneco Inc (USA)
Georges Bataille began his working life as a tenant farmer at Le Plessis-Belleville north of Paris. In 1927 he set up a small agricultural machinery repair workshop which was converted into the 'Poclain' production unit three years later. The limited slew TU excavator emerged in 1951 and in 1961 the company's first full slew machine was introduced, the TY45. The HC300 appeared in 1968 with weights up to 48 t and shovel buckets up to 3.2 m³ (4¼ cu yd), and was

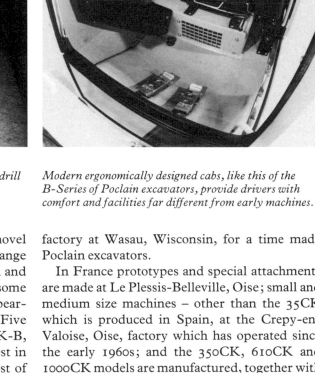

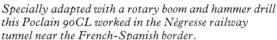

Specially adapted with a rotary boom and hammer drill this Poclain 90CL worked in the Négresse railway tunnel near the French-Spanish border.

Modern ergonomically designed cabs, like this of the B-Series of Poclain excavators, provide drivers with comfort and facilities far different from early machines.

followed in 1971 by the 137 t EC1000 with shovel buckets up to 8.7 m³ (11⅜ cu yd). A revised range of excavators appeared in 1974, re-designed and with up-dated hydraulic circuitry; in turn some of these were further amended with the appearance of 'B' series machines from 1980. Five models were introduced in 1982: the 70CK-B, 85CK-B, 125CK-B, 350CK and the smallest in the range, the 5 t 35CK. The current largest of twelve basic models is the 1000CK Series II, dating in its original form from 1975, and now weighing 190 t with shovel buckets up to 17 m³ (22 cu yd).

In 1966 Poclain joined Potain to form Potain-Poclain-Matériel (PPM). In the 1970s a range of hydraulically driven PPM draglines were produced at Montceau-les-Mines near Lyon in central France; now mobile cranes are made. In 1976 Tenneco became a major shareholder in Poclain, its J. I. Case division assumed responsibility for Poclain machines, and its Drott

factory at Wasau, Wisconsin, for a time made Poclain excavators.

In France prototypes and special attachments are made at Le Plessis-Belleville, Oise; small and medium size machines – other than the 35CK which is produced in Spain, at the Crepy-en-Valoise, Oise, factory which has operated since the early 1960s; and the 350CK, 610CK and 1000CK models are manufactured, together with machinery and welding undertaken for other factories, at the Carvin factory, Pas-de-Calais, opened in 1969. Poclain has subsidiaries in Spain and Mexico (both became operational in 1971) and its excavators are made under licence in Brazil, South Korea, India, Iran and Yugoslavia. From about 1965 until 1983 they were made under licence in Japan; from 1978 to 1980 in Czechoslovakia; and in 1982 production ceased at the Poclain factory in Buenos Aires, Argentina, and at the SMT factory in Tournai, Belgium, which had opened in 1972 and where large

excavators had been built. As well as excavators, Poclain also makes specialised hydraulic equipment in France and Eire.

RICHIER Nouvelle Industrielle Richier SA Richier was founded in 1929 by P. Richier to make concrete mixers; by the early 1960s a wide range of construction equipment was produced including cable 'Nordest' and hydraulic 'Oleomat' excavators. In 1972 Ford acquired a controlling interest in Richier, the range of products was reduced and excavators were marketed under the Ford name. Seven years later Ford sold its excavator manufacturing interests to Sambron SA which established Nouvelle Industrielle Richier SA (NIRSA) to manufacture Richier products and Distribution Internationale Richier SA (DIRSA) to distribute them. In 1980 the manufacture of excavators was moved from Charleville to L'Horme near St Etienne which is supplied with sub-assemblies from a factory at Sedan in the Ardennes.

Poclain 90P earthmoving beside the liner France.

Poclain's headquarters at Le Plessis-Belleville, 40 km (25 miles) north of Paris.

Sambron relinquished control of NIRSA and DIRSA in 1981; the company was operated by a manager appointed by the courts until it passed into the hands of a group of French industrialists in 1983. Bergerat Monnoyeur took control in 1984. Currently four basic crawler hydraulic excavators are made, H43, 45, 48 and 52; the first three have 'P' wheeled versions. Weights range from 11 to 30 t. Other products include rollers and a grader.

YUMBO   Yumbo SA

Production of hydraulic excavators started at Chauffailles north-west of Lyon, in 1954 when the Bruneri brothers ceded their patent rights on the first fully hydraulic excavator to SICAM, the predecessor company of Yumbo. From 1970 to 1982 the company was under the control of International Harvester, and between 1980 and 1981 the Eder M-815 was marketed as the IH620W. Now the company is an independent French manufacturer and assembly takes place at Genas, 10 km (6 miles) south east of Lyon – at a factory established in 1962 after a move from nearby Vaise. Welding, machining and part fabrication take place at Chauffailles. The present 600 Series of excavators which recently replaced the 3900 Series consists of four basic crawler machines, two of which are available in wheeled versions. Weights range from 12 t to the largest machine introduced in 1979, the 650 HD at 30 t. Three models of Yumbo excavators are made in Mexico.

INDIA

HEC   Heavy Engineering Corporation Ltd
Indian Government
This undertaking, as well as manufacturing steel mill and other heavy equipment, produces at Ranchi, Bihar, three makes of excavator under licence. Soviet EKG-4.6 shovels are manufactured; since an agreement reached in 1979, Rapier W2000 walking draglines are produced; and, following an agreement made in 1981 covering a part of B-E's line of electric mining shovels, the first of a batch of seven 195-BIs was shipped in 1982.

HIND (HIND-MARION VIKRAM)
Hindustan Motors Ltd
The company was established as a car manufacturer in 1942 at Port Okha, in the old Kathiawar state; in 1946 what was to become its main factory was opened at Uttarpara (now Hindmotor) in West Bengal. At Uttarpara in 1957 Hind's heavy engineering division was formed which now makes excavators, cranes and presses and does fabrication and machining work. Production of Marion excavators under licence began with 93 and 101-M models in 1961, the 111-M was added in 1978. In 1979 another licence was agreed, and from 1980 to the end of 1983 forty-five Demag hydraulic excavators were made: H36s are working at Lohardaga, Bihar, and Kasharsada, Belgaum, in bauxite mines, and the first H51 went in 1983 to the Manuguru coal mine, Andhra Pradesh. Production of H71 and H121 models is planned to start soon. Hind's other products include cars, trucks and a range of construction equipment.

L & T POCLAIN   Larsen and Toubro Ltd
Two Danish engineers, H. Holck-Larsen and S. K. Toubro founded L & T in Bombay in 1938: current products range from electronic controls to agricultural equipment. Manufacture of Poclain excavators began in 1975 at Byatarayanapura on the Bangalore-Hyderabad road. LC80 and LY80 models were originally made; the first machine going to the Kiriburu iron ore mines, Orisa. In 1977 the 90CK and 90P models were added, and in 1978 the 300CK; all three are currently produced. A total of some 500 units of all models have been made to date.

TATA (TELCO)   Tata Engineering and Locomotive Co Ltd
Boilers, and later complete locomotives, were amongst the products made at Jamshedpur after Tata Sons purchased an old East Indian Railway workshop in 1945; now Tata is a major producer of commercial vehicles. Excavator production results from a licence from P & H between 1962 and 1976; four P & H-derived basic cable models are made at Jamshedpur, the largest being the 1055B/BLC of 3 m³ (4 cu yd) capacity as a shovel or backhoe and 3.5 m³ (4½ cu yd) as dragline. Some 1,500 excavators have been made since 1962 and exported to countries including Spain, the Netherlands, Romania and Singapore.

INDONESIA

KOMATSU   Trakindo
Komatsu have recently made an agreement with

Trakindo for the assembly/joint-manufacture of Komatsu products which probably includes hydraulic excavators.

MITSUBISHI   p.t. Triguna Utama

A licence has recently been given to p.t. Triguna Utama for the assembly/joint-manufacture of Mitsubishi hydraulic excavators.

SUMITOMO   p.t. Traktor Nusantara

Sumitomo has recently agreed to the assembly/joint-manufacture of Sumitomo hydraulic excavators by p.t. Traktor Nusantara.

IRAN

P & H   Machine Sazi Arak

In recent years this company has manufactured P & H crawler draglines at Arak.

POCLAIN   Hepco Company

Models from the old range of Poclain excavators are produced by Hepco at Tehran. Production of models from the current range of machines is now starting.

ITALY

BENATI   Benati SpA
Gruppo Industriale BEN

In 1887 Andrea Benati started to manufacture ploughs of wood and iron. In the 1930s a range of agricultural equipment was being made by the company he founded; now Benati produce wheel loaders and backhoe loaders as well as excavators. The first excavator was the 8 t hydraulic MAX80 introduced in the 1950s and by 1969 the 23 t MAX 220 was in production. Manufacturing licences were granted in 1968 to Guria of Bilbao, Spain, and in 1969 to Tortone of Córdoba, Argentina. The size of machine subsequently increased through the 54 t BEN 525 in 1975 to the 96 t BEN 910 HDB in 1981 capable of handling buckets up to 6 m³ (7⅞ cu yd). At the 'Benati' factory in Imola, Bologna, thirteen basic CSB models of crawler hydraulic machines are made, the smallest being 12 t in weight; ten of these are available in long crawler (LCB) and six in heavy duty (HDB) versions. Three models of wheeled excavators are made at the 'Mater' factory. Benati also produces an 'Off-road 416' backhoe mounted on a Mercedes-Benz Unimog truck chassis, 125 and 145 'Railroad' machines, the IM14 telescoping boom excavator and special machines for applications in iron and steel works.

BENFRA   Benfra SoA

The Bendini and Frascaroli company was established in Bologna in 1947 to convert agricultural tractors for industrial uses. In 1956 the company moved to the present site at Modena, came under the ownership of stockholders and, until 1970 when its present name was adopted, was known as Bendini, Frascaroli & Co, SpA. The first Benfra hydraulic excavator was a wheeled machine introduced in 1961; by 1972 five crawler and three wheeled models were offered. Currently crawler 4, 5, 6, 8, 10 and 12 Ch and wheeled 4, 5 and 7 Rd models are made, the largest weighing 36 t. The Benfra line also includes backhoe loaders, wheel loaders and attachments for tractors, excavators and fork lift trucks.

COSMOTER   Cosmoter SpA

Cosmoter of Nogara, Verona, was established in 1969 and mainly manufactures excavators. The present range includes the 14 t BAT12RS wheeled backhoe and the crawler mounted BAT14CS, 18C and 22C backhoes with the 25 t BAT25C as backhoe or shovel.

DALLA BONA

From Brescia in northern Italy comes this range of small excavators mounted on single, unpowered, axles and designed for towing. The range includes the Major 2R and Super Mondial, both under 5 t weight. Dalla Bona machines are particularly used in agricultural operations and models are available with their own engines or utilise a tractor power take-off.

EUROMACH   Euromach snc

Three climbing-type hydraulic backhoes – the 2.8 t Buggy, 4.1 t Tigrotto and 5.05 t Tigrone – are manufactured at Montichiari, Brescia. A related company, Cormach, makes a range of agricultural implements, designed to be towed and powered by tractors, which can be equipped as small hydraulic backhoes.

FAI   FAI SpA

The company's first excavator, the FAI 1000, appeared in 1964 when FAI commenced operations; it was a tracked machine available as backhoe or shovel. The current range from the Noventa Vicentina factory, Vicenza, is composed of four basic crawler models, two of which are made in wheeled versions. Introduced in 1981, the present largest model weighs 21 t, but a larger

31 t 2000 model was discontinued in 1980. Backhoe loaders and skid steer loaders are also produced.

FIATALLIS   Fiatallis Europe SpA

Fiatallis was created in 1974 when Fiat, founded in Italy in 1899 to manufacture cars, and Allis-Chalmers, an American company dating back to 1847 when millstones were made, merged their construction equipment divisions. Fiat had been involved in the construction equipment field since 1946, and Allis-Chalmers had made steam shovels at Milwaukee, Wisconsin, in 1905. Both parent companies now have a very wide range of products – Fiat being particularly concerned with transportation equipment, Allis-Chalmers with machines relating to the production and use of power. Fiatallis excavators are the successors to machines made by Simit which merged with Fiat in 1972. The current 'FE' range, which replaced the previous 'S' range, is made at Grugliasco, near Turin, and consists of eight basic crawler models, two available in wheeled versions. Weights range from the 12 t FE12 to the 44 t FE40L with shovel bucket capacities up to 3 m³ (4 cu yd).

FMC LINK-BELT   FMC Corporation (USA)

At the Milan factory eight basic models of crawler dragline are made ranging from the 18 t LS-68 to the 76 t LS-338: two are also available as backhoes. Cranes are also produced here.

HYDROMAC   River SpA

Hydromac SpA was created in 1965 when the partnership of Carlo and Mario Bruneri was dissolved, and it subsequently came under the control of Ferruccio and Giampiero Bruneri, sons of Carlo. The Bruneri partnership had started in Turin in 1930 and in 1951 the brothers took out a patent for the first fully hydraulic, full slew, excavator. Hydromac excavators were made at San Mauro, near Turin, with part manufacture at another factory at Trino Vercelese.

The 'Superdigger' range in the late 1970s included six crawler and two wheeled backhoes. The renowned car designer Pininfarina, was involved in the design of the 30 t H180 and 34 t H200; the H180 had a unique ergonomically designed 'space age' capsule cab and striking decals. The company closed at the end of 1983 but River SpA, a mainly financial organisation,

bought the manufacturing rights. Production of the H95E Super, HG95, H120hd and H170, with overall weights ranging from 17 to 29 t, recommenced in 1984 at factories near Turin and Cúneo in north-west Italy.

KAMO   Moro SpA

Moro's Kamo 3X is a climbing type excavator made at Pordenone, near Venice. The machine weighs approximately 4.5 t and has bucket capacities up to 0.23 m³ ($\frac{1}{4}$ cu yd). Moro also manufactures small agricultural and industrial machines and pumps.

LALTESI   Laltesi Excavatori SpA

Luigi Laltesi founded the company in 1956 and since the first excavator was made, the small L30 crawler backhoe, some four thousand units have been produced at Alseno, Piacenza. Seven hydraulic backhoes are currently manufactured from the 8.1 t 111 LCL to the 28 t 551S; a prototype has been made of the 38 t L200. Laltesi has been associated with Macchine Agricole Industriali Automezzi SpA and Compagnia Generale Trattori SpA.

MACMOTER   Macmoter SpA

Based at Modigliana, south east of Bologna, is Macmoter – owned by Luis Haringer and Dr Ernst Baumgartner. It has made excavators since 1979 and the range has been designed by the Gamma Design Group of Turin. In 1984 four crawler models were in production – the 4.6 t M3, M5, M7 and 9 t M9 together with the M5R and M9R wheeled versions; possibly in 1985 the M11 at about 11 t will appear. Much of the company's output is marketed by the West German firm, Zeppelin, and in the UK from 1984 two Macmoter models are marketed as the O & K 4.6 t 32B and 10 t 58B. Skid steer loaders and backhoe attachments are also produced.

PMI   Padana Macchine Industriali SpA

At Carpaneto, Piacenza, PMI produces a range of five hydraulic backhoes from the 9.6 t 450B-CP to the 28 t 1500, plus the 2200 available as a backhoe or shovel with weights up to 43 t.

ROCK   Farben Industrial Development SpA

Rock crawler hydraulic backhoes have been made near Turin since 1978. Currently the Rock 100, 130, 150 and 200 models are made with weights ranging from 16 to 29 t. Farben ID mainly manufactures machinery for asbestos-cement pipe and sheet production.

JAPAN

AIRMAN Hokuetsu Industries Co Ltd
Hokuetsu was founded in 1938 and has two factories in Niigata Prefecture near Tokyo. The company is a major manufacturer of air compressors, and other products include diesel generators and engine welders. Mini backhoes have been made since 1981 and the five current crawler models range from 1 to 4.5 t with standard bucket capacities of up to 0.2 m³ ($\frac{1}{4}$ cu yd).

HINOMOTO (COMBAC) Toyosha Co Ltd
Toyosha of Kadoma city, Osaka, manufactured the 'Combac' range of mini excavators including the CR12, CR15 and the 4 t CR18 with a bucket capacity of 0.18 m³ ($\frac{1}{4}$ cu yd). However, excavator production ceased during 1983.

HITACHI Hitachi Construction Machinery Co Ltd
Hitachi Ltd is major shareholder
The Hitachi company was founded in 1910 and the construction equipment division, with products including excavators, cranes and tunnel boring equipment, became a separate entity in 1970. Varied products of the group as a whole range from turnkey nuclear power plants to subminiature semiconductors. The manufacture of electric quarry shovels commenced in 1939 and ten years later the 'U' series of mechanically driven universal cable machines was started with the UO5. Production of 'U' series machines was terminated in the late 1970s and these have been superseded by the KH series of five hydraulically driven draglines; the largest is the KH180-2. In 1965 Hitachi introduced the crawler mounted UHO3 backhoe, the first hydraulic excavator designed and produced in Japan. The UH series now consists of hydraulic excavators with weights ranging from 6.8 to 157 t. From 1974 to 1979 the largest was the UH30 with weights up to 77 t until the UH801 appeared at double that weight. Excavators are manufactured at the Tsuchiura factory, 75 km (47 miles) north east of Tokyo, which was opened in 1966. In 1981 about

IHI have been making hydraulic excavators in Japan since 1971.

7,100 hydraulic excavators were produced; by then a total of over 60,000 machines had been made. Hitachi excavators are produced under licence in South Korea, and an agreement was reached in 1983 for three models of Deere-engined excavators, assembled by Hitachi, to be marketed as Deere machines in North America from 1984.

IHI (ISHIKO)   Ishikawajima-Harima Heavy Industries Co Ltd

In 1960 Ishikawajima Shipyard, founded in Tokyo in 1853, and Harima Shipbuilding and Engineering Co Ltd, established in Aioi in 1907, merged to form IHI; products range from ships and jet engines to large industrial complexes. Cable excavator production dates from 1952 when a licence with Koehring came into being; this was terminated in 1981. Hydraulic excavators were first made by IHI in 1971. The company now markets in the 'IS' range, five mini backhoes made for IHI by Takeuchi; eight basic models of hydraulic backhoes up to the IS-310 with its 1.4 m³ ($1\frac{7}{8}$ cu yd) bucket; two draglines, the K250 and K400A; and the K1000 which is a 1.9 m³ ($2\frac{1}{2}$ cu yd) universal machine. The 3.1 m³ (4 cu yd) 1495 dragline at 116 t, and 1405 backhoe and shovel version with weights up to 95 t, are now only made to special order. Some IHI hydraulic excavator models will be marketed in North America from 1984 as Koehring machines.

IWAFUJI (HYDREX)   Iwati Fuji Industrial Co Ltd

Iwati Fuji manufacture mini hydraulic backhoes. Since 1982 two models have been altered by Atlas to match European requirements and are marketed as the Atlas 250 and 350 machines.

JSW (NIKKO)   The Japan Steel Works Ltd Nihon-Seiko-Sho

The Japan Steel Works was established at Muroran in 1907 by the Hikkaido Colliery and Steamship Co Ltd and two English firms – Sir W. G. Armstrong Whitworth & Co Ltd and Vickers Sons & Maxim Ltd. It now produces steel forgings, castings and plates as well as industrial machinery. Since 1964 excavators have been manufactured by JSW, up to 1979, under a licence with O & K; production started with the RH5. Recently the company has developed its own range of machines in conjunction with GK Industrial Design Associates. The NC220, introduced in 1984, is a JSW design; but earlier models, the BH70 and BH45 were modified versions of the O & K RH-6 and RH-4 respectively. Prior to the appearance of the NC220 there was a range of five crawler backhoes and shovels coming from the Fuchu City factory at Tokyo which was established in 1941; these ranged from the BH30 at 6.2 t to the BH110 II at 27 t. Also made was the 52 t LH300 shovel, apparently derived from the O & K RH25. JSW makes its own hydraulic pumps and motors at Yokohama.

KATO (HY-DIG)   Kato Works Co Ltd

Kato's origins go back to the founding of Kato Ironworks in 1895. Subsequently manufacture of internal-combustion engined locomotives has given way to products such as cranes, earth drills, and street sweepers. The first excavator was made in 1967, the company being only the second in Japan to design its own machines. In 1983 nine basic models were produced at the Higashioi factory in Tokyo and, since 1980, also at the Gunma factory in Ohta City; models ranged from the 4.5 t HD-180G to the HD-1880 SE with weights up to 43 t. In 1984 the HD-2500 was introduced.

KOBELCO (P & H KOBELCO YUTANI KOBELCO)   Kobe Steel Ltd

Kobe Steel, founded in 1905, as well as being a major crude steel manufacturer, includes aluminium and copper items, industrial machinery and construction and surface mining equipment amongst its products. The company produced Japan's first electric shovel in 1930, a 50K with a 1.5 m³ (2 cu yd) bucket and followed this in 1931 with the 3 m³ (4 cu yd) 120K and in 1934 with the 4 m³ ($5\frac{1}{4}$ cu yd) 200K. The K series was produced until 1954, then in 1955 a licensing agreement began with P & H and the first P & H designed excavator, the 255A, was made in 1956. In 1959 excavator production was moved from the Iwaya factory to the Okubo factory, at Akashi City near Kobe. During the 1960s P & H electric mining shovels, such as the 1400 and 1900, and hydraulic excavators starting with the H208 in 1967, were introduced. After 1972 P & H-derived hydraulic excavators were replaced by the R series including Kobe-designed models, and the R907 and R935 based on Liebherr models made under a

MANUFACTURERS AND THEIR PRODUCTS

First electric mining shovels to be built in Japan were
Kobe 50K models introduced in 1930.

licence from 1972 to 1979. In the 1970s also the
range of electric mining shovels was extended;
the 2800 was introduced in 1976. In 1980 Kobe's
own complete line of hydraulic excavators appeared and since then some 14,000 units have
been built; this hydraulic K series was updated in
1983 and currently comprises nine basic models
ranging up to the 130 t K975 with a 7.5 m³
(9⅞ cu yd) shovel bucket introduced in 1984.
Kobe's present cable excavator line consists of
four crawler draglines up to 0.8 m³ (1 cu yd)
capacity, and six electric mining shovels up to the
2800XP with a 23 m³ (30 cu yd) bucket.

In 1983 Kobe purchased Yutani and discontinued most of its excavator production; since
then Kobelco hydraulic excavators up to the
0.7 m³ (1⅞ cu yd) K907C size have been made at
the re-equipped, and now highly automated,
former Yutani factory at Hiroshima. Larger
hydraulic models and crawler draglines are pro-

duced at the Okubo factory constructed in 1942,
and electric mining shovels in a building opened
in the mid-1970s at the Takasago factory acquired in 1953 in Kobe City. Kobe obtained 10
per cent of P & H stock in 1981 to become the
company's largest single shareholder; the licensing agreement on all equipment other than
electric mining shovels was replaced by a partnership, and in 1983 a further agreement allowed
for the gradual transfer of manufacture of P & H
construction equipment to Kobe in Japan.
KOMATSU (KOMATSU-BUCYRUS)
Komatsu Ltd
Takeuchi Mining Company established the
Komatsu Ironworks in Komatsu City in 1917 to
manufacture tools for its own use and mining
equipment. Komatsu Ltd was created as a separate entity in 1921 and soon developed a wide
range of products including agricultural tractors
and, in the early 1940s, bulldozers. Now products include construction and earthmoving
equipment; industrial vehicles and steel castings.
Excavator manufacture dates from 1963 when

Excavating at the foot of Mt Fuji in Japan in about 1978 was this Kobe ('Kobelco' since 1980) 0.3 m³ (³⁄₈ cu yd) R903.

Komatsu, Mitsui and Bucyrus-Erie created an organisation allowing Komatsu to produce and market in Japan excavators incorporating B-E technology. Hydraulic backhoes mounted on Komatsu bulldozer undercarriages were made including the 10, 12, 15 and 20-HT models; by the 1970s the 195-B and 280-B electric mining shovels had been added. Since 1980 Komatsu has marketed its own PC and PW range of hydraulic excavators: these consist of five crawler minis; two basic models of small wheeled backhoes; seven basic models of medium size crawler backhoes; the larger 40 t PC400-1 and PC650-1, weighing up to 69 t which went into production in 1982 – both of which are available as backhoe or shovel; and the 160 t PC1500-1 shovel. In 1984 re-designed Series III versions of some PC models appeared. Machines smaller than the PC200 are assembled at the main Awazu factory in Komatsu City, Ishikawa prefecture, which dates from 1938 and has made excavators since 1974 – production of some models being transferred from the Osaka factory in 1981. The PC200 and larger models are built at the Osaka factory in Hirakata City between Osaka and Kyoto, which opened in 1952 and which has made excavators since 1963. Individual components are manufactured in other factories such

as engines at Oyama and hydraulic equipment at Kawasaki. Since 1983 Komatsu hydraulic excavators have been assembled/part-manufactured in Malaysia and probably also in Indonesia.

KUBOTA   Kubota Ltd

From a foundry opened in Osaka in 1890 by Gonshiro Kubota the Kubota company has grown into a manufacturer of, for example, small and medium sized tractors, marine and other engines, and building materials. The construction equipment field was entered in 1956 with production of mobile cranes, and excavators were added in 1966 with the commencement of a licence to make Atlas machines. In 1974 Kubota developed its own small excavator, the KH-1 and two years later the Atlas licence was terminated. Eight basic models of hydraulic backhoes are now produced ranging from the 1 t KH-5 to the 5.25 t KH-28; all are crawler mounted except the wheeled KH-16W introduced in 1982. Excavators and other small items of construction equipment are made at Hirakata in a factory opened in 1962 and extended in 1979. Since 1976 Kubota markets under its name larger Hitachi-made excavators and Hitachi markets under its name smaller Kubota-made machines. Daewoo now make a Kubota mini backhoe, and since 1983 JCB has produced a mini backhoe using Kubota components.

MINEX   Handozer Industry Co Ltd

Handozer, founded in 1963, manufactures back-hoe loaders, crawler carriers, hydraulic breakers and at least three mini backhoe models – the 2.82t 212T, the 3.3t 315T and the 4.48t 520T.

MITSUBISHI   Mitsubishi Heavy Industries Ltd

The present name of the company dates from 1934 but its origins can be found in the Nagasaki Forge of Takugawa Shogunate – founded in 1857, and Tsukumo Shokai – founded in 1870. Products now are as diverse as steel structures, aircraft, chemical plants and consumer goods. Excavator manufacture at the Akashi factory, established in 1960, was initiated under a 1960–77 licence from Yumbo and in 1961 Mitsubishi became the first hydraulic excavator manufacturer in Japan. However, since 1972 Mitsubishi has produced machines to its own designs and these have been updated in 1976 and

1979. Some twelve basic MS models of excavator are now in production: the smallest are the MSO3M and MSO4M mini backhoes at 2.8 t and 3.8 t respectively; the largest series produced model is the MS580 introduced in 1981, with weights up to 65 t. A prototype 165 t MS1600 has been made. Mitsubishi excavators are now assembled/part-manufactured in Indonesia.

MITSUI Mitsui Engineering and Shipbuilding Co Ltd

This company – shipbuilder and manufacturer of steel structures, heavy machinery and chemical plants – signed an agreement in 1981 with Bucyrus-Erie to allow manufacture of B-E electric mining shovels at its Tamano City factory. At the beginning of this century Mitsui distributed Bucyrus dredgers and from 1963 to 1980 the company had co-operated with B-E and Komatsu in Komatsu-Bucyrus KK which gave Komatsu the right to produce cable and hydraulic excavators using B-E technology. In 1982 the first two units resulting from the new licence, 295-BII shovels destined for Brazil, were ordered from Mitsui.

NISSAN Nissan Kizai Co Ltd

Nissan make a range of crawler mini excavators from 2.8 t to 4.54 t including the N-10SS, N-21SS, N-31SS, N-41SS and N-50SS. Recently introduced is the wheeled NW-300 at 2.96 t and capable of a road speed of 14.7 km/h ($9\frac{1}{10}$ mph); this machine was also marketed by Yutani as the 120 MH. Since 1983 Schaeff have marketed three Nissan models, one modified.

SUMITOMO (SUMITOMO-LINK BELT
SUMITOMO-MARION)
Sumitomo Heavy Industries Ltd

The business origins of the Sumitomo group can be traced to Kyoto in the early part of the seventeenth century when the Sumitomo family began copper mining and trading; diversification took place towards the end of the nineteenth century and now Sumitomo's activities encompass metals, shipbuilding, electronics, banking and trading. Resulting from a licence granted in 1963 with Link-Belt, seven basic models of crawler cable machines are produced, modified to meet Japanese requirements: the LS-78 and LS-408 are available with backhoe, shovel or dragline equipment; the LS-418 can be either backhoe or dragline; and the LS-108, 128, 518

and 528 are draglines only. Still based on FMC Link-Belt machines, but with greater Sumitomo design input, eight hydraulic excavators are produced up to the 45 t LS-5800J. Cable machines are made in the Nagoya (Ohbu) factory where the construction machinery plant dates from 1964, hydraulic ones are produced at the Chiba factory. Sumitomo hydraulic excavators are also now assembled/part-manufactured in Indonesia. At the Ehime factory Sumitomo-Marion electric mining shovels have been made since a licence with Marion was agreed in 1970.

TAKEUCHI (PEL-JOB) Takeuchi
Manufacturing Co Ltd

Founded in 1963 this company markets its small hydraulic backhoes itself as 'Pel-Job' in France and 'Takeuchi-Job' in the USA through subsidiaries in Venissieux and Atlanta, Georgia, respectively. It also specialises in producing excavators to be marketed by others: in the UK as the Priestman Mini Mustang range and in Japan as IHI and Yanmar models. Takeuchi's first mini excavator was made in 1971, and exporting started in 1978 with Taiwan and Singapore as additional destinations. By 1981 a range from 1.1 to 4.9 t had been created and by the following year the factory at Murakami had a capacity of at least 400 units per month and a total of over 20,000 excavators had been made. Six models are currently produced, prefixed by TB (for Taiwan, USA, Europe), MM (UK), EX (Singapore), IS (IHI-Japan) and YB (Yanmar-Japan). A 6.5 t model is being developed. Takeuchi also manufactures a small grader and industrial agitators.

YANMAR Yanmar Diesel Engine Co Ltd

Yamaoka Hatsudoki Kasakusho, the company which was the fore-runner of Yanmar, was established by Magokichi Yamaoka in Osaka in 1912. As well as being a specialist manufacturer of diesel engines, Yanmar also makes small items of construction equipment, tractors and small boats. Mini hydraulic excavators have been made since 1970; from 1981 at the Seirei Fukuoka factory. Seven crawler backhoes range from 1.1 t to 4.5 t and a 2.8 t wheeled backhoe, the YB1200W, appears to use a similar undercarriage to those on the Yutani 120MH and the Nissan NW-300. Some excavators marketed by Yanmar are manufactured by Takeuchi.

YUTANI Yutani Heavy Industries Ltd

Major shareholders, Marubini Corporation and Mitsubishi Heavy Industries Ltd

Yutani Komuten was founded by Eiji Yutani in 1917 at Osaka; from this developed the Yutani company which built construction equipment since 1939 at the Gion factory in Hiroshima and, since its opening in 1969, at the Numata factory where larger sizes of excavator were made. Many Yutani hydraulic excavators were based on Poclain designs, such as the TY45A and LC80S, and manufactured under a licence agreed in 1962. In recent recent years the company developed its own range and marketed the wheeled 2.96 t YS120MH and the YS1400-3 with weights up to 41 t. Yutani has also manufactured a wide range of special hydraulic attachments for logging, drilling and ship applications. During 1983 the company was bought by Kobe Steel and production of Yutani excavators was terminated except for two models now made at Kobe's Okubo factory.

MALAYSIA

KOMATSU   United Motor Works

PC120 and PC200 Komatsu hydraulic excavators have been made since 1983 under a licensing agreement for assembly/joint-manufacture with UMW.

MEXICO

FMC LINK-BELT

At Queretaro LS-68, LS-98 and LS-108B draglines, and the LS-98PL backhoe, are made under licence.

POCLAIN   Poclain Mexicana Poclain SA (France)

Production of excavators at the company's Mexican subsidiary at Naucalpan de Juarez commenced in 1971. The TY45, LC80, TCS, LY2P and 90CK are currently manufactured.

YUMBO   Maquinaria Hidraulica Mexicana SA de CV

Maquimex, founded in 1969, produces three models of Yumbo excavator: the 640HD and 630L crawler machines, and the 3965 wheel mounted one similar to the 630L.

NETHERLANDS

JOHN DEERE   De Rotterdamsche Droogdok Maatschappij

RDM and Deere have jointly designed a 12–16 t range of wheeled and crawler hydraulic excavators. These will be produced by RDM in Rotterdam and marketed from 1984 under the Deere name. RDM also make military equipment, heavy hydraulic machinery and steam turbines.

VERMEER   Vermeer-Holland BV

In the middle of the nineteenth century the Vermeer family emigrated from the Netherlands to Iowa. In 1945 a descendant, Gary Vermeer, began manufacturing small agricultural machines and around 1956 the first Vermeer Ditchers were exported to the Netherlands. From 1959 assembly took place in that country and shortly after a Dutch, but American financed, manufacturing company was established. A separate Dutch company was created in 1964 and in 1978 the Vermeer family sold their interests. Vermeer now specialises in manufacturing trenchers and earthboring machines at Hoofddorp, south west of Amsterdam. However, three basic models of mini hydraulic backhoes are also produced: the crawler mounted 1.4 t Little Dinky, 2.6 t Dinky Digger and 4 t Big Dinky. A 2.1 t wheeled version of the Dinky Digger has been recently developed for gravedigging.

NORWAY

BRØYT A/S

The Søyland brothers, Ingebret trained as a carpenter and Kristian as a car mechanic, established Bryne Treog Jernindustri in 1949 in Ingebret's carpentry shop created from their mother's redundant hen house; this was the beginning of Brøyt at Bryne in the Jæren coastal area in the south-west of the country. Farm products were made until an order was received for an excavator to dig ditches locally; the result was the appearance in 1952 of the 4K, a cable operated backhoe towed and powered by a tractor. Before long hydraulics were applied and the X2, introduced in 1958, embodied the features found in most subsequent Brøyt machines – an undriven undercarriage, available as backhoe or shovel, and special front-end geometry. Current Brøyt models are the 13 t X21 Series II backhoe and the 33 t X41 shovel – both first manufactured in 1978, together with the more recent 47 t X50 equipped with a 4 m³ ($5\frac{1}{4}$ cu yd) shovel bucket. In

1981 Åkerman took a 20 per cent holding in Brøyt; collaboration was initiated and production of the crawler mounted X21TL was discontinued as well as marketing of the X12 – an Eder M-815 marketed under the Brøyt name since 1980.

HYMAS   A. S. Hymas

Hymas was founded in 1904 and specialises in the production of backhoe-loaders. Tractor and truck mounted cranes are also made; together with dental chairs! The 7.5 t 82R backhoe of 0.27 m³ ($\frac{3}{8}$ cu yd) capacity is the single excavator model manufactured.

POLAND
WARYNSKI (BUMAR ZAKLADY)
Kombinat Maszyn Budowlanych Bumar.
Warszawskie Zaklady Maszyn Budowlanych
Warynski
Polish Government

In 1958 the hydraulic excavator industry was established with production of the 10 t KUH-251. Universal cable machines were made under a licence from Menck in the 1960s and British Priestman Cub excavators were produced in the same decade – as the crawler KM-251 commencing in 1962 and the Star truck mounted KS-251 in 1967.

Currently two factories in Warsaw, one established in 1833, make hydraulic excavators under the Warynski and Bumar names. Previously the name 'Zaklady' has been used, and 'Unikop' was a former manufacturer. Models in the present range of eleven, with weights up to 54 t – the M500H, have been introduced since 1969. Nine models are available as shovel or backhoe, six of these being crawler mounted, two are wheeled,

The Polish 1611 Brawal backhoe is capable of some of the movements of a telescoping boom excavator.

Promex build ESE 8001 shovels in Romania, probably based on the Soviet EKG-81.

and the 407B is mounted on a Star truck. The 415 is a crawler hydraulically driven dragline and the 1011 is a crawler hydraulically driven universal machine; both were introduced in 1983. The K Series of five models are Polish designed but use British Leyland engines and Hamworthy hydraulic systems built under licence. The M250H and M500H are Menck-Koehring models made under licence since 1972 and the 1611 Brawal is a recent innovative machine.

ROMANIA
PROMEX   IUG PROGRESUL
Romanian Government

Established in 1924, as a Franco-Romanian enterprise to repair locomotives and ships, the Progresul factory at Brăila has made excavators since 1956. Around 50 per cent of the excavators made are exported. Currently seven basic models of hydraulic machines range from the 11 t truck mounted A402 backhoe, to the 144 t crawler SC7001 backhoe or shovel with bucket capacities up to 8 m³ (10½ cu yd); models with wheel or crawler mountings and long crawler or heavy duty versions are available. Three models can be converted to hydraulically driven draglines including the DH3602 with a telescoping boom and capacities up to 2.5 m³ (3¼ cu yd). Also manufactured by Progresul are the S1202T telescoping boom excavator and the ESE 8001 electric

mining shovel, probably based on the Soviet EKG 8I, with capacities up to 12 m³ (15¾ cu yd). Nearly a hundred Liebherr 981 excavators were made under licence in Romania in the late 1970s.

SOUTH KOREA
DAEWOO   Daewoo Heavy Industries Ltd
Daewoo was established in 1937 as a state-run company but in 1976 it merged with Korea Machinery Industries Ltd and is now a company with shareholders. Machine tools, diesel engines, rolling stock and construction equipment are amongst the group's products. Hydraulic excavators have been made since 1978 at Incheon under a licence agreed with Hitachi the previous year. Three crawler backhoes are produced: the DH04-2, 07-3 and 09, and by far the largest number made so far are 19 t DH07s. There is also available a wheeled version of the DH04, and a 51 t crawler DH20 model is being promoted. In addition the recently introduced DH02 is a 6.2 t crawler backhoe designed by Kubota. Excavators are exported particularly to the Middle East and other parts of the Far East.
HALLA POCLAIN   Korea Heavy Industries and Construction Co Ltd
Two Halla excavators, the wheeled 60P and crawler 90CK are produced by KHIC of Seoul under a licence from Poclain which dates from about 1977. A wider section of the Poclain range was previously made, up to the 220CK, when the manufacturer was known as Hyundai International Inc.

Hitachi excavators being built under licence at the Daewoo factory, Incheon, South Korea.

This Çukurova 750A was employed on a sewage project for Izmir in Turkey in 1982.

EXCAVATORS

SPAIN

GURIA   Guria S Coop Ltda
Caja Laboral Popular

Industrias Guria SCI, established in 1961, was
engaged in shipbuilding and manufacturing public works machinery. In 1980 Guria S Coop was
created as a separate company making excavators
and backhoe loaders at Alto de Arretxe, Irun,
Guipúzcoe. Guria is an industrial co-operative,
in turn part of the CLP co-operative group
created in 1956 with diverse products including
household electrical goods, machine tools and
bicycles. Excavators have been made by Guria
since 1968, at first under licence from Benati:
exporting commenced in 1977. Seven models of
hydraulic excavators are now produced ranging
from 11 to 46 t. These include three wheeled and
four crawler backhoes. The largest crawler machine, the 545, is also available in HD version as a
shovel.

MTM   La Maquinista Terrestre Y Maritima,
SA

Included amongst MTMs excavators in recent
years have been the 14 t 90-C crawler backhoe
and its 15 t 90-R wheeled version, together with
the 22 t 130-R backhoe with four wheel drive.

POCLAIN   Poclain SA (France)

In the 1960s TUSA marketed Yumbo excavators; from 1971 the TUSA factory in Zaragoza
has operated as a Poclain subsidiary. Currently the
35CK, 60CK/P, 75CK/P and 90CK/P models
are made. Some 280 units of the 0.5 m³ ($\frac{5}{8}$ cu yd)
t11 crawler machine were built for the Spanish
market during 1972–81. TUSA also makes a
range of industrial storage and conveying
equipment.

SWEDEN

Åkerman   Åkermans Verkstad AB

Lars Petter Åkerman bought a foundry at Eslöv
in 1889 to establish a company which began by
manufacturing castings for such things as steam
engines and pumps – even producing items as
varied as cast-iron church windows and, in 1904,
part of a submarine. In 1925 construction machinery was added to the company's products
and in 1938 Åkerman's manager, Bernt-Lorenz
Åkesson, developed the first excavator which
appeared the following year. This 300 model was
an 8 t universal crawler machine with a 0.3 m³ ($\frac{3}{8}$

*The Åkerman H25 was produced between 1971 and
1974 and it has been in its C version since 1979.*

cu yd) bucket. Åkerman's first hydraulic excavator was introduced in 1965, the H11 weighing up to 19 t, and the last cable excavators were
built four years later. Åkerman gave licences to
Rimas of Ringköbing in Denmark (c. 1950–63)
and to Varkaus of Varkaus in eastern Finland
(c. 1950–67). Both companies produced mostly
200-D models; Rimas also made the 300 and 375
and Varkaus the 300, 375, 475, 575 and 751. In
1972, before Kockum-Landsverk excavator production ceased in 1973, Åkerman took over the
remaining sales and service of their former
Swedish competitor. In 1981 the company took a
20 per cent shareholding in Brøyt, and also in that
year Åkerman acquired the Hein-Werner
Corporation's Construction Equipment Division at Waukesha, Wisconsin. Over 11,000 Åkerman excavators have been produced. In 1983
there were six basic models with weights from 14
to 57 t, and in 1984 the 7.8 t H3, with offset boom
facility, was added. The full range is manufactured at Eslöv, and hydraulic components have
been made at Lund since 1980 in the factory of
the former cable excavator manufacturer,
Åsbrink. At Waukesha, Åkerman Hein-Werner
H16D, H14B and H10 models are made together

with some original H-W machines.

VG Vaggerydsgrävare AB

The VG110RL made at Vaggaryd is a 3.4 t crawler mounted hydraulic backhoe with a standard bucket capacity of 0.1 m³ ($\frac{1}{8}$ cu yd). Its prime uses are for peat cutting, moor and forest ditching and small excavation tasks.

SWITZERLAND

MENZIE MUCK Menzie AG

Menzie was established in 1933 and initially manufactured agricultural implements. Since 1966 climbing-type excavators have been produced at the Widnau factory. Three versions each of the 3000 and 5000 models of 'Climbing Hoe' are made with weights up to 13 t; there is also the 5000 Mobile which can be fitted with a third driven wheel. The Climbing Hoe was introduced into North America in 1975. Recently Menzie has begun to market a range of five mini excavators incorporating Japanese components; the smallest is the 10S.

TURKEY

ÇUKUROVA Çukurova İthalât ve İhracat T A Ş

Çukurova Holding AŞ

Çukurova was established in 1923 as a textile company and takes its name from the Çukurova plain, a cotton growing area bordering the Mediterranean in Southern Turkey. Diversification has taken place so that now the group markets and services construction and agricultural equipment, manufactures chemicals and plastics and has, amongst others, textile and banking interests. At Izmir, on the Aegean coast, Çukurova made a prototype hydraulic backhoe in 1979 and the wheeled 750 model has been in production since 1980. By mid-1983 thirty-five units had been made; the A series being replaced by the B series at the beginning of that year with an increase in weight from 14 to 15 t.

UK

HYMAC NEI Thompson Hymac

Northern Engineering Industries Group

Hymac's origins date back to 1827 when the Butc Iron works was established. Ten years later it amalgamated with the Union Iron Company to form Rhymney Iron Company, in the town of

that name in South Wales where ore was smelted with coal from its own collieries. Iron making ceased in 1891 but coal continued to be mined and in 1920 the site was acquired by the Powell Duffryn Steam Coal Co. In 1946 Powell Duffryn and International Combustion Ltd formed Rhymney Engineering Co Ltd to produce mining equipment. However, in 1961 a lack of demand for such products caused the company to diversify and it obtained a licence from the Hydraulic Machinery Co of Butler, Wisconsin, to make Hy-Hoe hydraulic excavators. The marketing of these 480 models with 270° slew was undertaken by companies within the Lehane, Mackenzie and Shand (LEMAND) Group; one of which was led by Peter Hamilton who played a significant role in introducing hydraulic excavators into the UK. At the end of 1963, after 318 units had been made, the 480 was replaced by the British-designed 0.5 m³ ($\frac{5}{8}$ cu yd) 580; versions of this machine were a prime product of Hymac until the early 1980s, but other models were later introduced, the largest of which was the 33 t 1290 which appeared in the early 1970s. In 1966 Rhymney Engineering became a wholly owned subsidiary of Powell Duffryn and two years later the excavator marketing companies were incorporated to form Hymac Ltd. In 1972 Hymac purchased Whitlock Brothers of Great Yeldham, Essex, founded in 1963: it then terminated production there of the Fuchs 703R made under licence as the Whitlock 50R. Between 1977 and 1980 Hymac marketed Demag excavators in the UK.

Hymac and the West German firm, Hanomag, were acquired by the IBH Group in 1980. Hanomag was established by Georg Egestorff in 1835 as an iron foundry and engineering works; locomotives, trucks and cars were amongst Hanomag's subsequent products. In 1920 construction equipment was first made and since 1970 the company exclusively manufactured this type of product. Originally Hanomag marketed hydraulic excavators manufactured by its then parent company, Rheinstahl Union Brückenbau; later it acquired the line of Hagelstein (HATRA) machines before producing its own modified versions of these in Hanover. Immediately prior to its takeover by IBH, Hanomag was owned during the 1970s by Massey-Ferguson during

EXCAVATORS

which time M-F excavators were made at
Hanover and, after the acquisition of Beltrami
excavators, at Aprilia in Italy. Also, certain Eder
models were marketed by Hanomag and M-F.

The 1983 Hymac range consisted of four basic
hydraulic excavators ranging from 12 to 23 t: the
crawler 580D/DS introduced in 1980 – Hymac
designed; the crawler 450E/ELC and its wheeled
W450E version – Hanomag designed; the crawler 201/LC and its wheeled W201 introduced in
1983 – designed jointly by Hymac and Hanomag;
and the wheeled W350D – Hanomag designed.
All were manufactured at Rhymney other than
the W350D which was still made at Hanover.

The IBH Group, formed in 1975, collapsed
in 1983 and the last excavator was made at
Rhymney at the end of that year. Early in 1984
Hymac was bought by NEI and production of all
Hymac models except the W350D was recommenced soon after at Ettingshall, near Wolver-
hampton, West Midlands. NEI's varied engineering products include many concerned with
power generation, transmission and distribution
– such as steam turbines, marine engines and
switchgear.

JCB   J. C. Bamford Excavators Ltd
In the 1930s Joseph Cyril Bamford worked in the
family business constructing agricultural equip-

*J. C. Bamford, holding his son – JCB's present
Chairman, and the company's workforce in 1947.*

*The 808 was the largest excavator made by JCB and its
rural setting is appropriate for a manufacturer which
started by making agricultural equipment.*

Priestman grab, built 1921, and the hundredth Priestman Cub, a dragline built 1933; preserved at Hull, Humberside.

ment and stationary engines at Uttoxeter, Staffordshire. He established his own firm in 1945, in a small rented garage in Uttoxeter, to produce purpose-built trailers for agricultural tractors. A move to other premises – stables – soon followed and in 1950 the company moved to its location beside the River Churnet at Rocester, 6.4 km (4 miles) to the north of Uttoxeter. This site, previously a cheese factory, pig farm and cattle market, was to be developed into JCB's modern and impressively landscaped factory. By the time of this move JCB had built hydraulic tipping trailers and produced its first hydraulic loader attached to a tractor; by the mid-1950s a backhoe tractor attachment had been made available and the line of JCB backhoe loaders was established. The company's first excavator appeared in 1965, based on a W & S Hopto design; from this JCB7 with a bucket capacity of 0.48 m³ ($\frac{5}{8}$ cu yd) soon up-rated to 0.57 m³ ($\frac{3}{4}$ cu yd) evolved the 7C in 1966 and the 8C and 8D in 1971. A further series evolved from the 6 and 6C introduced in 1966 – the 6D in 1968 and the 7B in 1969; and the 5C, also introduced in 1969, provided the basis for the present 800 series which was started in 1972. All three standard current backhoes, weighing overall up to 21 t,

first appeared in 1982; the mini 802 was introduced in 1983 and incorporates Kubota components. The 808, weighing up to 24 t, has recently been deleted from the range after some ten years in production. JCB, which is still a private family owned company, as well as manufacturing backhoe loaders, makes wheel loaders, rough terrain forklift trucks and hydraulic power packs.

MANIDIG   Manitou (Site Lift) Ltd
Manitou BF SA (France)
Manitou's 3.25 t Manidig 325 crawler mounted mini backhoe was originally introduced in 1981 when, as the B15-X, it was manufactured by a company established by Ian Beresford at Redditch, Hereford and Worcester. In 1982 the manufacturing and marketing rights were acquired by Manitou and changes made before re-launching at the end of that year under its new name. The Manidig was at first still built at Redditch but production has since been transferred to Verwood, Dorset. The parent company manufactures rough terrain forklift trucks in France.

Priestman's line of excavators developed from the ditcher sketched on an envelope by Sydney Priestman in 1920.

POWERFAB   Powerfab Ltd

The climbing-type 0.8 t Powerfab 360W backhoe designed by David John, a former Hymac employee, is manufactured by his company at Tredegar, Gwent. It was introduced in 1982 and is probably the smallest full slew excavator in the world. Powerfab also makes a limited slew 125W backhoe.

PRIESTMAN   Priestman Brothers Ltd
Acrow Group

After serving his apprenticeship in shipbuilding at Hull, Humberside, and locomotive engineering at Gateshead, Tyne and Wear, William Dent Priestman worked for a short time as an engineer at Newcastle upon Tyne, also in Tyne and Wear. In 1870 his father bought the Holderness Foundry in Hull and set him up in business, with Richard Sizer, making replacement parts for windmills and machinery for paint and sugar manufacture. The partnership was dissolved after only two years, but William's brother, Samuel, came into the business in 1873 and soon after this the company became Priestman Brothers. A winch was designed and manufactured in 1875 for use in an abortive attempt to raise sunken treasure off the Spanish coast. Priestman's first steam grab for use on land was sold in 1877 and the next year the company's first grab dredger was put into operation. Oil engines were made in the 1880s and eighteen Priestman crane grabs were used on construction of the Manchester Ship Canal; crane grabs developed as the company's main product in the early years of this century. Sydney H. Priestman, son of Samuel, designed in 1920 a wheel mounted ditching grab towed and powered by a crawler tractor; the No 1 Grab Ditcher was produced in 1921. In 1924 the first dragline was made; this and the following five units were half-track machines, but later ones were full track types. A backacter was added to the growing range in 1926. The No 5 model was developed in 1932 into the 7 t Cub with a 0.19 m³ ($\frac{3}{4}$ cu yd) bucket which became a fully universal machine and the first of many series of Priestman excavators to carry the name of a wild animal. The Mustang range of fully hydraulic excavators was initiated by the wheeled Mustang 90 introduced in 1967. Also during the 1960s, Priestman marketed Yumbo excavators. The current range of crawler excavators made at Marfleet on the eastern outskirts of Hull, to which Priestman moved between 1950 and 1958, includes the unique VC15 and VC20 machines, two Mustang backhoes with respective weights up to 12 and 16 t, and three hydraulically driven Lion draglines with overall bucket capacities up to 3 m³ (4 cu yd). In addition Priestman markets four Mini Mustang backhoes made by Takeuchi. Priestman also makes marine cranes, grabs and grab dredging cranes and a subsidiary produces Taperex crossed roller bearing slewing rings – developed in 1956 and first used on a Tiger excavator. In 1969 Priestman joined the Steel Group Ltd which in 1972 became part of the Acrow Group – a group which had its origins in 1936 and consisted of companies making such products as cranes, panel and Bailey bridges, scaffolding and formwork. The Acrow Group collapsed during 1984.

RANSOMES & RAPIER (NCK-RAPIER)
Ransomes & Rapier plc Central & Sheerwood plc
J. A. Ransome, R. J. Ransome and R. C. Rapier left a local engineering works in 1869 to form a partnership and create their own establishment on the banks of the River Orwell at Ipswich, Suffolk. Ransomes and Rapier at first specialised in producing railway equipment – from wedges and the wooden pegs that held chairs to sleepers (the oak sawdust was used at nearby Lowestoft for fish curing and in return kippers were brought to the factory at Christmas!) to complete locomotives such as the first one to be used in China, in the 1870s. R & R later manufactured a wide range of products including cranes, turntables and water control gates, and developed as a 'one-off' specialist; it was only in the 1960s that general engineering was phased out. The first excavators were made in 1914 when two 36 t cranes for the New South Wales railway in Australia were equipped as shovels to enable them to load coal onto locomotives. Twelve years elapsed before the next excavators were made, under a licence from Marion between 1925 and 1930. Three types of steam or electric powered R & R-Marion excavators were made, Type 7, 460 and 480. After the agreement was terminated R & R was permitted to use the information it had gained on Marion designs for subsequent Rapier excavators, including thirteen 5360 Series stripping shovels, each weighing over 600 t, built between 1934 and 1942 for English iron ore mines. In 1939 R & R entered the walking dragline field and fifty-six machines were built up to 1964.

In 1958 Newton Chambers & Co Ltd took over R & R. This company had been founded in 1792 when George Newton and Thomas Chambers started their partnership to smelt and cast iron hardwares in Sheffield, South Yorkshire. Within two years they obtained the coal and ironstone mining rights of the Thorncliffe Valley at Chapeltown, north of the city, and developed there a major industrial complex later to include blast furnaces, a heavy iron foundry, and a coal tar distilling plant from which by-products such as germicides were produced. This company's initial range of NCH excavators was produced between 1935 and 1947 under a licence from P & H. In 1947 a further licence was arranged, with Koehring, and the first NCK excavator – the 304 model – was manufactured. This licence ran until 1973 with NCK, and later NCK-Rapier excavators being produced at Chapeltown until 1982.

In the early 1960s Rhodes & Halmshaw of Ossett, West Yorkshire, which built most of the excavator cabs, was acquired and Koehring was for a time a major shareholder in the R & R part of the company. Also in this decade there was a rationalisation of the NCK and Rapier ranges with original Rapier models eventually being phased out in favour of NCK ones. During the late 1960s and early 1970s hydraulic excavators were produced. The company was purchased by the present owners in 1972.

R & R now produce thirteen models of crawler cable excavators; the largest is the hydraulically driven Olympus HC170 dragline with bucket capacities up to 3.8 m³ (5 cu yd). A new range of walking draglines was introduced in 1976 and now comprises seven models. A total of eleven units of only the 7.6–13 m³ (10–17 cu yd) W700 and the 26–34 m³ (34–45 cu yd) W2000 had been made at Ipswich, plus a W2000 made under licence at Ranchi, India, up to the end of 1983. All crawler excavator assembly is now undertaken at Ipswich, but components and part assemblies are still produced at Thorncliffe. R & R also manufactures cranes, pumps, and truck concrete mixers. The Central and Sheerwood Group has financial, printing, publishing and distribution interests as well.

Ransomes & Rapier made thirteen 5360 type stripping shovels from 1934 to 1942 for UK iron ore mines; a 100-B is in the foreground.

One of two Rapier W700s at the Godkin coal mine, near Heanor, Derbyshire in 1983.

RUSTON-BUCYRUS   Ruston-Bucyrus Ltd
Shareholders – Bucyrus-Erie Company (USA)
& General Electric Company Ltd

A chance remark – that he knew of a 'nice little
business' for sale in Lincoln – by a Sheffield,
South Yorkshire, barber in 1856 to his client
resulted in Joseph Ruston visiting Lincoln and
buying a major interest in the millwrights and
general smiths' business of Burton and Proctor.
Ruston, Burton & Proctor was thus established in
1857, making portable steam engines for agricultural use, and soon thereafter a very wide range of
heavy engineering products was added. Burton
left the partnership in 1857 and James Proctor
retired in 1864, although his name was retained
when the company became Ruston Proctor Ltd
in 1889. In 1875 the first excavator was produced
to the design of James Dunbar whose patent had
been obtained the previous year. By the end of
the nineteenth century a railroad shovel had been
developed and early in the twentieth the
company's first electric shovel and full slew
machine appeared. In 1910 the business of
Whitaker and Sons of Horsforth, near Leeds,
West Yorkshire – a company which had produced the first full slew excavator in 1884 – was
bought, and eight years later Richard Hornsby &
Sons of Grantham, Lincolnshire, was absorbed
and Ruston & Hornsby created. Seaman and
Hornsby had been founded as agricultural black-
smiths in 1815; following Seaman's retirement in
1828 Richard Hornsby and his sons had developed the company which went public in 1877.
Portable steam engines were followed by pioneering development of heavy oil engines and
caterpillar tracks.

By 1923 the R & H product line had developed
to the extent that the No. 300 dragline was
introduced with a 6.1 m³ (8 cu yd) bucket, and
the popular 0.38 m³ (½ cu yd) No. 4 universal
crawler model appeared in 1926. In 1930 Ruston-
Bucyrus was established, jointly owned by
Bucyrus-Erie which had operational control, and
Ruston & Hornsby which relinquished to R-B its
excavator business. R & H became part of
English Electric in the late 1960s, which in turn
became AEI before the present owners GEC.
B-E designed machines soon replaced R & H
models. Walking dragline production commenced with the 5W in 1939 and the largest R-B

*A chance meeting in a barbers' shop in 1856 led Joseph
Ruston to create what was to become one of the longest-
established excavator manufacturers, Ruston-Bucyrus.*

*Ruston-Bucyrus machines at the port of Seaham on the
Durham coast; an 1100 Supercrane derived from the
61-RB and a 38-RB Heavy Duty – both equipped as
grabs.*

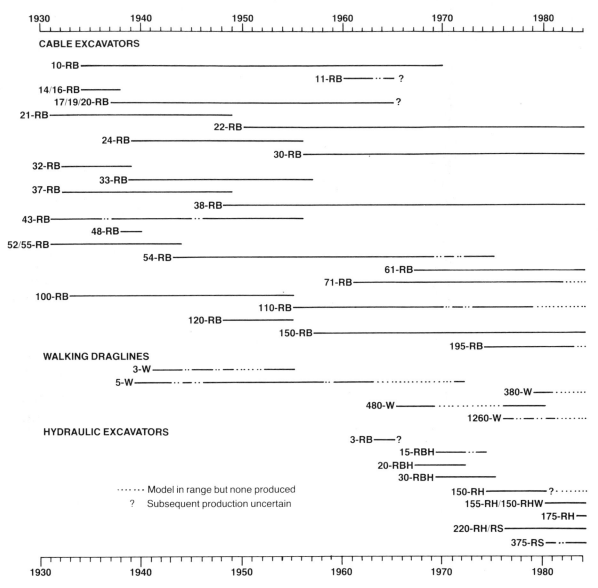

Ruston-Bucyrus excavator models.

electric mining shovel, the 195-RB, went into production in 1974. Although R-B produced its own 3-RB hydraulic excavator from 1963 to 1964, and B-E designed 15, 20 and 30-RBH models from 1967 to 1974, it was not until 1976 that an R-B designed series of machines was launched with the 150-RH. For a time from 1969 the 20-RBH had also been built in Australia by English Electric Diesels Australia Ltd. R-B's largest excavator, the 50 m³ (65 cu yd) 1550-W 'Big Geordie', designed by B-E and constructed in association with F. H. Lloyd & Co Ltd of

Wednesbury, West Midlands – which, since the 1930s, has supplied R-B with such things as steel castings – was put to work in Northumberland also in 1969.

In 1983 R-B's range consisted of eight basic crawler mounted cable excavators extending to the 342 t 195-B, with overall bucket capacities up to 12.2 m³ (16 cu yd); the modular 380-W walking dragline with capacities up to 12.2 m³ (16 cu yd) and the 1260-W up to 30 m³ (40 cu yd); and one wheeled, the 150-RHW, and three crawler, the 155-RH, 175-RH and 220-RH backhoes, together with the crawler 220-RS shovel and 375-RS shovel which was introduced in

Ruston No.6 engaged on coastal work.

1981. The weights of hydraulic machines ranged from 15 to 52 t, but production of all hydraulic models was suspended at the end of 1983. R-B also makes marine, crawler and truck-mounted cranes.

SMALLEY   Smalley Excavators Ltd

Richard Smalley, who devised the climbing-type backhoe in 1959, established the company at Osbournby, Lincolnshire, in 1962. In 1983 it ceased to be a solely family-owned business and assembly was moved to Bourne, Lincolnshire, with fabrication of parts now being undertaken by sub-contractors. The 5 model is also assembled at Christchurch, New Zealand, by Southern Cross Engineering. Seven models of small hydraulic excavators are produced with weights ranging from 1.84 to 8.5 t. These include the 5 series 3 climbing-type backhoe; the 470, a four wheeled self-propelled 0.35 m³ (½ cu yd) backhoe incorporating Poclain hydraulics; and the 808 crawler backhoe introduced in 1982 with low ground-bearing pressure for peat drainage and mosquito control work. A limited slew machine is also made for special applications and in 1984 the crawler 2.8 t 5T was introduced.

SMITH   NEI Cranes Ltd

Northern Engineering Industries Group

Jeremiah Balmforth, David Smith and Jeremiah Booth formed a partnership in 1820 to operate as millwrights and 'furnishers' of woollen mills. They established their factory at Rodley alongside the Leeds–Liverpool canal, on the outskirts of Leeds, West Yorkshire. Soon gas fittings, stone cutting machines and winches were made and in 1840 hand-operated cranes were added. In 1847 Jeremiah Booth sold his interest to the other parties and established his own crane works in Rodley; in the early 1980s, when the Booth & Smith companies were under common NEI ownership, they again amalgamated. Thomas Smith, who had succeeded his father David, in 1861 bought out the share of William Balmforth

129

30-RBs being built in 1980 at Ruston-Bucyrus' Lincoln factory, Lincolnshire.

who had succeeded his father Jeremiah. The company was then operated by the Smith family – with Thomas' three sons entering into a partnership in 1902 and a limited company being formed in 1918 – until 1939 when Thomas W. Ward Ltd obtained a 75 per cent controlling interest in Thomas Smith & Sons (Rodley) Ltd. The 1860s saw the introduction of a range of steam cranes and the first Smith excavator appeared in 1887 when a shovel attachment was fitted to a 3 t rail mounted steam crane. During the building of the Manchester Ship Canal, Smith cranes were used just as cranes and also fitted with Whitaker shovel attachments; for some years Smith operated a factory at nearby Bramley to meet the demand. However, with the exception of a few 'Jubb Trencher' backacter machines, based on cranes and incorporating a patent bought by Smith in 1902 off a contract foreman in the Manchester area, regular excavator manufacture ceased until 1930. In that year two universal crawler excavators were produced – the 0.38 m³ (½ cu yd) 'Half Yard' which became the 4-14 and the 0.25 m³ (⅓ cu yd) 'Third Yard' which became the 3-12 and 2-10; to be

followed four years later by a 0.19 m³ (¼ cu yd) No 7 model. By 1939 Smith had sold 461 of these excavators. From then on Smith developed a range of cable excavators up to 1.1 m³ (1½ cu yd) as shovels and 1.9 m³ (2½ cu yd) as draglines, as well as making hydraulic truck, cable crawler and truck, dockside, rail and rough terrain cranes. In 1976 two 'Eurocrane' hydraulically driven draglines were launched and in 1979 the 0.57 m³ (¾ cu yd) 25E mechanically driven dragline appeared, designed to appeal to developing countries because of its ease of operation and maintenance. Although a Super 10 prototype hydraulic excavator was made at the beginning of the 1960s, Smiths only production of hydraulic excavators was between 1968 and 1973 when French Tractem machines were built. Smith excavators have been manufactured at factories other than at Rodley; for example, the 0.57 m³ (¾ cu yd) 21 introduced in 1946 was made by other companies in the Ward Group – Butters at Glasgow, Strathclyde, and (with the 26)

The Jubb Trencher was made by Smith at the beginning of the twentieth century.

Marshall at Gainsborough, Lincolnshire – and by W. S. Thomas and Taylor at Benoni, South Africa, in 1950–55. The No 8 was made at Keighley, West Yorkshire, by John Smith Cranes. Smith have been a part of the NEI Group since 1978, and the 1983 range of Smith excavators consisted of six basic models of draglines ranging from the E1400 with bucket capacities up to 0.5 m³ ($\frac{5}{8}$ cu yd) to the E4000L/W up to 1.9 m³ ($2\frac{1}{2}$ cu yd).

USA

ÅKERMAN HEIN-WERNER   Åkerman HW Inc
Åkermans Verkstad AB (Sweden)

Hein-Werner was founded in 1919 and produced its first excavator in 1961, the truck mounted T10, followed in 1962 by the crawler C10, based on a design of the Hydraulic Machinery Co Inc of Butler, Wisconsin. In 1962 Hydraulic Machinery began to manufacture 'Hy-Hoe' excavators itself, hence Hein-Werner designed and made its own models although the major part of the company's activities has, certainly recently, been concerned with other products such as hydraulic tools and components. Nevertheless, up to 1981 some two thousand H-W excavators were made, including the Rotograder telescoping boom model.

Immediately prior to the acquisition of H-W's Construction Equipment Division in 1981 by Åkerman, C-12B, C-14B, C-24 and C-28 models of crawler hydraulic backhoes were being made, from 18 to 49 t, at Waukesha, Wisconsin. Production of the Åkerman Hein-Werner H16C and H14 commenced in 1982, followed by the H10 in 1983; with some original H-W machines still being made.

AMERICAN   American Hoist and Derrick Company

Although Oliver T. Crosby and Frank J. Johnson established The Franklin Manufacturing Company in 1882 in St Paul, Minnesota, essentially as a machinery repair business, the following year the company started to produce hand winches and horse powered hoists and changed its name to the American Manufacturing Company; in

*President Roosevelt boarded an 86 t Bucyrus railroad
shovel when visiting the Panama Canal construction
work in 1908.*

1892, it received its present name. By the end of
the nineteenth century wooden derricks and
locomotive cranes were American products and
in 1904 an excavator was devised by converting a
rail mounted 'log loader' into a railroad steam
shovel. In 1905 this became the American Railroad Ditcher. Similarly, in 1923 three locomotive
cranes were mounted on crawlers to make the
company's first crawler machines; in 1928 this
design became the base for the American Gopher
Shovel Crane. By the end of the 1930s, as well as
winches and cranes, there was a line of American
excavators equipped with buckets up to 1.5 m³
(2 cu yd) – all petrol engined since 1932. Between
1931 and 1937 Dominion Hoist & Shovel Co Ltd
of Montreal, Quebec, commenced excavator
manufacture with the 300, a Gopher made under
licence. In 1955 American acquired the Wayne
Crane Division of American Steel Dredge based
in Ft Wayne, Indiana; this company formed in

1906, produced its first steam shovel in 1910 and
wheeled and crawler excavators were introduced
in 1946 and 1949 respectively. The Wayne acquisition provided American with 0.38 and 0.57 m³
($\frac{1}{2}$ and $\frac{3}{4}$ cu yd) models which remained in production until 1972. American excavators are pro-
duced at St Paul and at Ft Wayne – to which
production from a factory at Duluth, Minnesota,
was transferred in 1982. Now five series of
crawler draglines are made with overall capacities
up to 11.5 m³ (15 cu yd), together with five
hydraulic backhoes ranging from the 19 t 185 to
the 78 t 780. American also manufactures a wide
variety of cranes, bucket wheel excavators,
winches and machinery for the waste management, fire protection and logging industries.

BADGER (HYDRO-SCOPIC)   Badger
Construction Equipment Co
Burro-Badger Corporation
Badger Machinery Company, incorporated in
1946, in Winona, Minnesota, developed a
'Hopto' tractor mounted backhoe and later hy-
draulic excavators. Between 1958 and 1978 Bad-
ger was a division of Warner & Swasey during

which time it produced W & S hydraulic back-hoes including the 94 t 1900, cranes, telescoping boom excavators and special vehicle chassis. JCB excavators were also developed from a W & S Hopto design during this period. Since 1978, when it came under its present ownership, Badger has continued production of the crawler Hopto 'eleven' series of backhoes, introduced towards the end of W & S ownership, with the 24–51 t 211, 311 and 411 models. Also two models of Hydro-Scopic telescoping boom excavators either truck or crawler mounted, and hydraulic cranes, are made.

BUCYRUS-ERIE   Bucyrus-Erie Corporation Daniel P. Eells of Cleveland, Ohio, was active in promoting what became the Ohio Central Railroad to carry coal from the south-west of the state, via Bucyrus, to the Great Lakes port of Toledo. He was instrumental in organising a group of people to purchase in 1880 a factory at Bucyrus – the location of many of the railway's operational headquarters – and formed the Bucyrus Foundry and Manufacturing Company to make railway and coal mining equipment. Two years later the Ohio Central Railroad ordered a 41 t steam shovel and the first Bucyrus excavator was made, Model No 1 of 'The Thompson Iron Steam Shovel and Derrick'. Dredge building commenced in 1883. By the end of 1894 171 shovels had been manufactured. In 1893 the company moved to its present location at South Milwaukee, Wisconsin, and in 1896 it became The Bucyrus Company.

A licensing agreement was made in 1900 for Bucyrus shovels to be made in Russia at St Petersburg; this lasted until 1910 but was extended by tacit agreement until the revolution in 1917. Also between 1904 and 1909 Bucyrus machines were made under licence in Toronto, Ontario. In 1910 the Vulcan Steam Shovel Co of Toledo, Ohio, which had been building steam shovels probably since about 1877 was acquired by Bucyrus and Bucyrus-Vulcan formed. In 1911, Bucyrus, Bucyrus-Vulcan and the Atlantic Equipment Company – makers of Atlantic Railroad shovels, joined together to create Bucyrus Company with an additional new factory at Evansville, Indiana. By 1912 Bucyrus had full slew, crawler and dragline machines in production. By the end of the 1920s large stripping shovels were in existence, the first diesel powered excavator had been produced, the last railroad shovel was shipped, and dredge production had greatly declined. In 1927 the Erie Steam Shovel Co of Erie, Pennsylvania, was acquired. Bucyrus-Erie with Ruston & Hornsby (a predecessor company of which had produced its first excavator seven years before Bucyrus), formed Ruston-Bucyrus in the UK in 1930. Walking draglines were added in 1931 with the purchase of the Monighan Manufacturing Company of Chicago, Illinois, and Bucyrus-Monighan was formally merged with B-E in 1946. The Milwaukee Hydraulics Corporation was taken over by B-E in 1948 and that company's Hydrocrane was developed into B-E's first hydraulic excavator; although it was not until 1965 that the forerunners of the present range appeared. Characteristic of the 1960s was the introduction of very large machines – ultimately the 107 m³ (140 cu yd) 3850-B stripping shovel and the 168 m³ (220 cu yd) 4250-W walking dragline.

In 1963 a joint venture, Komatsu-Bucyrus KK, was created and cable and hydraulic excavators using B-E technology, including B-E designed electric mining shovels, were subsequently made in Japan; this agreement was terminated in 1981. Also in 1963 a B-E affiliate was established in Mexico but production did not commence until 1970; in 1965 a joint venture between a B-E subsidiary and FNV in Brazil started excavator production; in 1974 there was part-manufacture of B-E walking draglines in Australia and South Africa by other companies; and in 1981 agreements were reached for Mitsui in Japan, and HEC in India, to produce electric mining shovels. B-E excavators have also been made in Chile, France, New Zealand and Sweden. Excavators were made by B-E at Guelph, Ontario, between 1955 and 1971. Since the closure of its Evansville, Indiana, factory in 1981, B-E has also closed the Erie, Pennsylvania, and Pocatello, Idaho (opened in 1974), factories and concentrated all excavator production at South Milwaukee. The range of excavators offered in 1983 – the widest of any manufacturer – comprised five models of hydraulic backhoe and a hydraulic mining shovel; four cable shovels; seven crawler draglines; three cable backhoes; five electric mining shovels; a mining size crawler

Still working in Alberta when photographed in 1983 was this 1936 B-E 950-B with counterbalanced hoist : the machine was retired in 1984.

Excavating Hyde Park railway station at Sydney, Australia, in 1925 with a Bucyrus Class 14 dragline.

One of nineteen Erie Bs shipped to the UK in 1919.

dragline; and thirteen walking draglines. Although four models of stripping shovel were also offered, none of this type of excavator has been made by B-E since 1969. From early 1984 B-E has reduced its range by deleting many construction-size machines including the 71-B and other smaller cable models, and the 300-H, 325-H and 350-H hydraulic backhoes. In 1981 B-E diversified by purchasing a company involved in the production of aerospace components, gears and transmission systems; as well as these in 1983 B-E manufactured a range of cranes, winches, drills, logging machines, backhoe-loaders, bucket wheel excavators and dragline buckets.

CASE DROTT   J. I. Case Company
Tenneco Inc

An improved 'Ground Hog' thresher, made in 1842 in Rochester, Wisconsin, by Jerome Increase Case was the first product of the Case company. He moved to Racine in 1844 – still the company's headquarters – and the J. I. Case Threshing Machine Co was incorporated in 1880. As well as threshing machines, steam and petrol engined traction engines were made; construction equipment appeared in 1912 with grader and steam roller production. From 1964 to 1967 Kern County Land Co owned Case, then it was purchased by Tenneco Inc.

In 1900 Ed A. Drott began work in a logging camp; eighteen years later his first sale of a Holt crawler tractor was to that industry. Ed Drott opened a garage in Butternut, Wisconsin, in 1920 to sell cars and Holt tractors. In 1925 he moved to Wausau, Wisconsin, and his Hi-Way Service Corporation began manufacturing attachments for tractors. Growth resulted in a move to Milwaukee, Wisconsin, but the company – Drott Manufacturing Company since 1945 – moved back to Wausau in 1948. Drott entered the hydraulic excavator field in 1962 by purchasing the Yumbo manufacturing rights but two years later produced its own Cruz-Air wheeled machine.

Drott was acquired by Tenneco in 1968 and became a division of Case. The first 'Case'

A Case Drott 40, with electro-magnet, in a scrap yard at Colorado Springs, Colorado.

excavator, the 880, was introduced in 1971. All current Case-Drott excavators are made at Wausau; these are the 40 Series E, 50 Series E and 880C crawler backhoes and the 40 and 45 Cruz Air wheeled machines with overall weights from 16 to 23 t. Up to 1983 the Poclain 220 was also built at Wausau and Case markets Poclain excavators under the Drott name in North America. Case also manufactures other items of construction equipment such as backhoe loaders and wheel and crawler bulldozers as well as agricultural machinery, cranes and forestry equipment. Tenneco operations cover, for example, the oil, natural gas and chemical industries.

CATERPILLAR   Caterpillar Tractor Company

Towards the end of the last century, Benjamin Holt of Stockton, and Daniel Best of San Leandro, provided steam traction engines to replace horses as the motive power for combined harvesters used on the soft soils of the San Joaquin Valley, California. In 1904 Benjamin Holt first fitted crawler tracks to one of his tractors; his expanding company moved to East Peoria, Illinois, in 1909; and the name 'Caterpillar' was registered as a trademark in 1910. C. L. Best, Daniel's son, who had started his own tractor manufacturing business in 1910 after his father had sold out to Benjamin Holt two years previously, produced his first crawler machine in 1913. Both companies merged in 1925 to form the Caterpillar Tractor Company. Now Caterpillar, which is a multi-national company with over 40,000 shareholders (none of whom owns as

much as 1 per cent of the total stock), operates world-wide and produces around 140 models of earthmoving, construction and materials handling machines. Caterpillar excavators have only been manufactured since 1972; the first model, the 225 backhoe, was followed by the 235 and 245 available as backhoe or shovel – all three US-designed, and the most recent, the 215 backhoe (in B version from 1984), was designed in Belgium. Overall weights range from 17 to 67 t. There is also a heavy duty version of the 215B, the 215BSA mounted on a 225 undercarriage; and from 1984 Caterpillar are marketing the 205, 211, 211LC, 213 and 213LC crawler models, and 206, 212, 214 and 224 wheeled models – all Eder machines. Excavators are made at Aurora, Illinois, in a factory established in 1958 and at the Gosselies factory in Belgium which opened in 1968. Caterpillar engines and parts of crawler undercarriages are supplied to other excavator manufacturers.

FMC LINK-BELT   FMC Corporation

In 1894 the Link-Belt Company of Chicago, Illinois, introduced a rail mounted coal handling clamshell; soon locomotive cranes, and later power transmission and materials handling equipment, were added. The first crawler excavator was introduced in 1922 and by the late 1930s machines with capacities up to 1.9 m³ ($2\frac{1}{2}$ cu yd) were available. In 1939 Link-Belt purchased the Speeder Machinery Company which had been established at Leon, Iowa, in 1919 to manufacture an unconventional light excavator, the 'Tumble Bug'. Speeder production moved via Fairfield to Cedar Rapids, Iowa in 1926 and by 1939 0.29–0.57 m³ ($\frac{3}{8}-\frac{3}{4}$ cu yd) models were made. Subsequently a combined range of Link-Belt Speeder excavators was made, all at Cedar Rapids from 1948. In 1968 FMC took control of the company. Now there are twelve cable crawler machines, all available as draglines and two as backhoes, with weights from 18 to 117 t. There are also seven models of crawler hydraulic backhoes, from 20 to 77 t; the first four hydraulic models were introduced in 1967 and the oldest model in the present range dates from 1973. Ten cable models and five hydraulic models are made at Cedar Rapids. FMC Link-Belt excavators are also made in Italy, Japan and Mexico. Until recently the

LS-418A cable machine was made in the company's factory in Caçapava, São Paulo, Brazil, and also many cable and hydraulic models used to be manufactured in another FMC Link-Belt factory at Woodstock, west of Toronto, Ontario. FMC Link-Belt also produces a wide range of cranes.

GRADALL   The Warner & Swasey Company
The Bendix Corporation

Two former machinists, Worcester Reed Warner and Ambrose Swasey, started up in business in 1880 initially to produce machine tools for the plumbing industry. Within a year they moved from Chicago, Illinois, to Cleveland, Ohio. Refracting telescopes were added to their line but it was the machine tool operation which was to grow into W & S's main product. However, since the 1940s W & S has diversified and now also produces textile machinery and electronic equipment. In 1945 W & S bought the manufacturing rights of a basic machine to be developed into the Gradall telescoping boom excavator. Early units were built in Cleveland but from 1950 onwards they have been made at New Philadelphia, Ohio. Currently there are four basic models of Gradall. Between 1958 and 1978 the Badger Machinery Co of Winona, Minnesota, was part of W & S, and hydraulic backhoes as well as Badger's Hydro-Scopic telescoping boom excavators were then W & S products. Also, in 1967, W & S acquired Sargent Engineering Co of Ft Dodge, Iowa, a former manufacturer of cable excavators since the 1920s; this company was disposed of in 1978.

INSLEY   Insley Manufacturing Corporation
L. Boren, R. Doermer and Narreagansett
Capital Corporation

Insley Iron Works was established at Indianapolis, Indiana, by William H. Insley in 1905 to manufacture fabricated structural steel; the company also soon developed concrete placing and re-handling equipment and later made cranes. The company produced its first excavator in 1924; of 0.28 m³ (⅜ cu yd) capacity, it was mounted on a Fordson tractor and had 180° slew. The following year the first of some 600 crawler mounted Model C machines appeared of the same capacity but with 210° slew. During the 1930s and 1940s Insley C excavators were produced in the UK under licence by Blaw Knox at

Worcester – Hereford and Worcester, Watford-Hertfordshire and Rochester-Kent. Insley's first full slew excavator, the Model R, was introduced in 1928 and the 'K' series, based on two models of universal cable machines, was in production in various forms from 1935 to 1971. The hydraulic market was entered in 1963 with the H-100 and the current series of six crawler backhoes range from the 15 t H-600 to the 59 t H-3500C. The Insley company, which was given its present title in 1936, was acquired by W. P. Elliot in 1943 and from 1969 to 1976 was part of the Desa Industries organisation before being sold to AMCA International Corporation. Insley was taken over by its present owners in 1979.

JOHN DEERE   Deere & Company
John Deere served his apprenticeship in Middleburg, Vermont, and in 1825 began a career as a journeyman blacksmith. He set up in business in Grand Detour, Illinois, in 1836, and the following year devised a steel plough to replace the cast iron ones then in use. In 1847 he opened a factory at Moline, Illinois, and Deere & Company was incorporated in 1868. The company still has its headquarters at Moline and as well as being a major producer of agricultural machinery makes, for example, equipment for the forestry, landscaping and materials-handling industries; since the 1950s it has expanded a line of construction equipment. Deere's first excavator, the crawler 690, was introduced in 1968 and the current B version weighing up to 18 t appeared in 1973. There is now a wheeled 690B for governmental use only. The other two current models are the A version of the 890, a machine which was first made in 1978, and the 990 weighing up to 43 t and introduced in 1982. For nearly three years in the late 1970s Atlas excavators were marketed as the wheeled Deere 34, 44 and 54 models and the crawler 45, 55 and 65 models. Hitachi are building in Japan three models of excavators fitted with Deere engines for marketing from 1984 under the Deere name in North America. Also from 1984 wheeled and crawler excavators, designed jointly by RDM and Deere, will be produced in the Netherlands and marketed under the Deere name in Europe, Africa and the Middle East.

KOEHRING (BANTAM LORAIN)
Koehring Cranes and Excavators Division

AMCA International Corporation (Canada)
A series of acquisitions of other excavator manufacturers has given rise to the present Koehring models. The original parent Koehring company, Koehring Machine Co – its name was changed to the Koehring Company in 1920 – was founded in Milwaukee, Wisconsin, in 1907 by Philip Koehring. He was an accountant who became an engineering designer and his company was created to manufacture concrete mixers; these with road making and repairing machines became major products. Excavator production commenced in 1922 with a petrol engined machine. In 1929–30 the company was associated with Insley in the National Equipment Corporation, and in 1952 the original company became the Koehring Division of Koehring Company. In 1959 Koehring acquired the former excavator manufacturer Stardrill-Keystone Co and closed the Beaver Falls, Pennsylvania, factory.

Schield Bantam Co became part of Koehring in 1963. Vern Schield, owner of a limestone quarry at Waverly, Iowa, had built his own truck mounted dragline in 1941. His brother, Wilbur, a vegetable wholesaler, joined him and production

The John Deere Company, famous for agricultural machinery, produced its first excavator in 1968.

The 440 Spanner with a capacity up to 1 m³ (1¼ cu yd) was one of four Koehring models available as draglines in 1983.

of draglines commenced at a 'factory' in the quarry in 1943. The brothers with two other partners established the Schield Bantam Co in 1945 and only truck mounted excavators were made until a crawler mounted model was introduced in 1954. Ten years later the first hydraulic excavator, the 450, was launched, by which time Schield Bantam was a part of Koehring. In 1969 came the first telescoping boom machine, the 725 Teleskoop, and the last cable excavator emerged in 1973. Between 1973 and 1976 some production was undertaken at Newton, Iowa.

The year following the Schield Bantam acquisition the Thew Shovel Co of Lorain, Ohio, was incorporated into Koehring; prior to 1920 it had

been the Thew Automatic Shovel Co Inc. In 1955 Thew had taken over the Byers excavator manufacturing company, established in 1873. Koehring's most recent aquisition took place in 1966 when the Menck company of Ellerau, West Germany, was added. Menck had produced cable excavators – including electric shovels and draglines over 3.8 m³ (5 cu yd) capacity – and two models of its then current hydraulic excavator line were absorbed into the Koehring line. Koehring itself was purchased by AMCA International in 1980; this group, founded in Canada

Work on a new runway at Detroit airport in 1983 being undertaken by a Koehring 1066 backhoe.

in 1882 as Dominion Bridge Company, is involved not only with construction equipment but also with a wide range of steel-based products for heavy industrial applications.

Koehring excavators were made in Victoria, Australia, by Armstrong Holland from 1945 to 1981; NCK machines were produced at Thorncliffe in the UK by Newton Chambers from 1947 to 1973; cable machines were made in Madrid, Spain, by Kynos from 1952 to 1974; 'Ishiko' machines were produced at Yokohama, Japan, by IHI from 1952 to 1981; and Koehring-Menck excavators have been made at Warsaw, Poland, by Warynski since 1972.

Hydraulic excavators up to 34 t are made at Waverly, above this weight at Milwaukee; since 1982 Bantam models have been marketed under the Koehring name. Cable excavators are made at the Chattanooga, Tennessee, factory which was established in 1952. The Lorain factory in Chattanooga finished manufacturing cable excavators in 1971, and the 48H hydraulic excavator in 1975. From 1955 to 1976 Koehring operated a factory at Brantford, Ontario, where the 205 and 605 were made. The present range consists of ten models of crawler hydraulic backhoes with two available as shovels, four crawler draglines, and two Bantam Teleskoop telescoping boom excavators – one truck mounted, the other truck or crawler mounted. From 1984 certain IHI hydraulic excavator models will be marketed in North America as Koehring machines.

LIEBHERR   Liebherr-America Inc
Liebherr International AG (Switzerland)
Liebherr's factory was established at Newport News, Virginia, in 1971. Currently three models of crawler hydraulic excavators are made, the R942, R962 and R982.

LITTLE GIANT   Little Giant Crane and Shovel Inc
The only excavator now made by this company, of Des Moines, Iowa, is the 34TX telescoping boom machine. This is available in either truck or crawler versions. Previously Little Giant made crawler and truck mounted cable excavators, and now also makes cranes and equipment carriers.

MANITOWOC   Manitowoc Engineering Co
The Manitowoc Company Inc
Two men employed in the shipbuilding industry, Elias Gunnell and Charles C. West, were in-

strumental in the purchase in 1902 of the Burger & Burger Shipyards. These repaired wooden ships on a meander core of the Manitowoc River at Manitowoc, Wisconsin. The Manitowoc Dry Dock Company was thus formed; in 1905 it launched its first steel vessel and in 1908 a nearby boiler works was bought and other products requiring heavy metal fabrication – such as machines for the pulp and paper industries (which are still made) – were added. In 1916 three of Manitowoc's principal shareholders had obtained an interest in a sand and gravel company which had eight Moore Speedcranes, presumably equipped with grabs, working in its material yard. These wheeled, steam powered, machines were made in Ft Wayne, Indiana, for Roy and Charles Moore of Chicago. Manitowoc began manufacturing Speedcranes in 1925 under the Moore patents; Roy Moore re-designed his machine as a universal, crawler mounted and petrol engined excavator, the 0.76 m³ (1 cu yd) Model 100 which was produced from 1925 to 1928. In 1928 Manitowoc took over sales as well as production. The first Manitowoc excavator, which was not just a re-designed Moore model, appeared in 1932. Subsequently, as well as producing excavators, Manitowoc has grown into a specialist builder of larger crawler cranes with ringer and tower attachments. Although a number of current models of cranes can be used as draglines, the 4600 Series 1 introduced in 1961 and available as shovel or dragline with capacities up to 6.1 m³ (8 cu yd), the 6400 dragline introduced in 1977 with capacities up to 11.5 m³ (15 cu yd) and the 3950D dragline introduced in 1984 with capacities up to 3.8 m³ (5 cu yd), are designed just as excavators. The company received its present title in 1952. In 1971 shipbuilding and repairing functions were re-established this time at Sturgeon Bay, Wisconsin – no large vessels have been constructed at Manitowoc since 1960.

MARION   Marion Power Shovel Division
Dresser Industries Inc
For a period in 1883 Henry M. Barnhart operated the steam shovel digging ballast from a gravel pit at Hepburn, west of Marion, Ohio, during the construction of a section of the Chicago and Atlantic Railroad. This experience led him to believe he could improve the

Marion's factory in the early 1900s.

machine's design. At that time Edward Huber's firm, the Huber Manufacturing Company, was making agricultural machinery in Marion itself and in part of the premises Henry Barnhart's cousin, George W. King, made hay carriers using castings made by the Huber company. Henry Barnhart approached Edward Huber to manufacture the steam shovel he had devised; George King joined them and the three men, together with associates, founded the Marion Steam Shovel Company in 1884. Three 'Barnhart Special' shovels were made in the Huber factory, but in 1885 the company obtained its own facilities on the site of the present Marion factory. To the Barnhart range of railroad shovels and wrecking cars were soon added ditchers, log loaders and dredges. Early in the twentieth century full slew, electric powered and crawler models appeared and Marion's 150 t Type 250 with its 2.7 m³ (3½ cu yd) bucket led the excavator industry into the large stripping shovel field in 1911. In 1939

the company's first walking dragline, the 7200, was produced and the following year the knee-action crowd system was applied to a 5561 stripping shovel. A range of construction-size excavators was phased out by the early 1970s but mining shovels, stripping shovels and walking draglines had developed rapidly. Marion produced the world's largest stripping shovel, the 138 m³ (180 cu yd) 6360, which started work in 1965, and the company's largest walking dragline, the 115 m³ (150 cu yd) 8950, became operational in 1973. The unique 204-M Superfront was developed in the late 1960s and early 1970s, and Marion's first hydraulic excavator, the 3560, was introduced in 1982. Now Marion's range is composed of five basic models of electric mining shovel (including the Superfront) and two dragline versions; the hydraulic 3560 as backhoe or shovel, and nine models of walking dragline.

Other than abroad under licensing agreements, all Marion excavators have been produced in the town of that name, the company's title changing to Marion Power Shovel in 1946. In 1910 former employees formed the Marion Shovel & Dredge Company which soon took over the remaining assets of former excavator manufacturer Osgood & MacNaughton. After a short period as Marion-Osgood the company became The Osgood Company. Another excavator manufacturer, the Fairbanks Company, operated in Marion from 1903 to 1925; then in 1926 the General Excavator Company was established in the factory formerly occupied by Fairbanks.

In 1949 Osgood and General merged, and in 1954 Marion acquired Osgood-General. In 1961 Marion took over the Quickway Crane and Shovel Co, established in 1922 in Denver, Colorado, and makers of mainly truck mounted excavators, but also of the cable-operated Chore Master telescoping boom attachment which became the Marion Grademaster; the Denver factory was closed in 1963.

The Marion railroad ditcher was an early full slew excavator; one of six Model 15s shipped to the Canadian Pacific Railway in 1914–19.

Boom walkways on 'Brutus', a Marion 8200 at Paintearth Mine in Alberta, indicate the enormous size of walking draglines.

Marion entered the hydraulic excavator field with the large 3560 in 1982, seen here during trials in Kentucky.

Marion's first licence was granted to Ransomes & Rapier of Ipswich, Suffolk, from 1925 to 1930, when Type 7, 460 and 480 models were made; also in the UK from 1959 to the mid-1970s Babcock & Wilcox produced a total of some two hundred and fifty 43, 45, 47, 93 and 101-M models at Dalmuir, near Glasgow, Strathclyde. In France from 1960 to 1975 Compagnie des Ateliers et Forges de la Loire (CAFL)/Creusot-Loire made a total of about a hundred 101 and 111-M models at St Chamond, near St Etienne, Loire; in India Hindustan Motors have made a total of over five hundred 93, 101 and 111-Ms since 1961 at Uttarpara (Hindustan), West Bengal; and since 1970 Sumitomo have made Marion electric mining shovels at Ehime in Japan. Limited production has included Babcock & Wilcox making three 191-M shovels in the mid-1960s; and in 1983–84 two 191-Ms were made by Buckau-Wolf at Grevenbroich, near Düsseldorf, West Germany. Part-manufacture of Marion excavators has also taken place in the UK, Australia, South Africa and Brazil.

Until 1927 the Marion company was a family business; thereafter it was a corporate body without 'local' controlling interests until local control was restored in 1944. A major interest in Marion was obtained by the Merritt-Chapman & Scott Corporation in 1954; by the early 1960s its subsidiary, the Universal Marion Corporation was the owner, and in 1965 Marion became a subsidiary of the Pittsburgh Coke and Chemical Company. The present owners took control in 1977. As well as excavators Marion produces blast hole drills and the Dresser group's construction and earthmoving equipment range – since 1983 incorporating from International Harvester the Hough line of machines and since 1984 Wabco products – includes dump trucks, coal haulers, bulldozers, wheel and crawler loaders, graders and cranes. Other products range from equipment for the petroleum and underground mining industries to diesel, petrol and natural gas engines.

NORTHWEST   Northwest Engineering
Company   Karin Corporation

Hartman-Greiling Company, founded in 1910, built boilers and did heavy structural engineering and machine shop jobbing. In 1918 its name was changed to Northwest Engineering Works and under the guidance of one of the company's originators, L. H. Barkhausen, who had also experience of contracting, development work on a crawler crane commenced. The first excavator, the 104, was in production from 1921 to 1929 and by the time of its introduction the company was known by its present name. Many early machines were designed by Paul Burke. Over the years Northwest developed the dual crowd for shovels, a crawler dragline with extended gantry for reducing boom compression, a system of control using engine power to throw heavy drum clutches, a hydraulic backhoe that could be converted to dragline and the 'DA Pullshovel' cable backhoe with hydraulically operated bucket wrist action. The 1.9 m³ (2½ cu yd) 80D has been in production since 1923. Northwest excavators have always been made at Green Bay,

Wisconsin, and hydraulic excavators were introduced in 1969. Recently the range comprised five basic cable machines; two available as dragline or DA Pullshovel, two as shovels and one as a dragline only. There were also three hydraulic backhoes introduced between 1976 and 1978, the largest being the 91 t 100-DH. Northwest also has produced lifting and magnet cranes and log loaders.

PAGE   Page Engineering Company

John W. Page devised a two-line dragline in 1904 and the Page company – incorporated in 1912 – after joint manufacture with the Monighan Machine Company from 1907 to about 1916 became a fully independent dragline manufacturer. Three and four legged rack systems for travel were used before the present cam walking device was introduced in 1935. Page has specialised in powering its walking draglines with its own large diesel engines. In the late 1960s and early 1970s a number of barge mounted clamshell dredges

A Page 400 Series dragline displaying the company's three-legged walking device.

One of five West German-built P & H 1200s operating at UK coal mines.

were manufactured. Currently there are eight electrically powered models of Page walking draglines; the largest being the 3,429 t 757 with bucket capacities up to 57 m³ (75 cu yd). The company, owned and operated by three Page daughters and their families, is based at a factory in McCook, near Chicago, Illinois. Page also supplies dragline buckets for use on other manufacturers' machines – over 60,000 to date.

P & H   Harnischfeger Corporation

Henry Harnischfeger served his apprenticeship as a locksmith in Germany before emigrating to the USA in 1872. After working in and around New York he moved in 1881 with another toolmaker, Maurice Weiss, to Milwaukee, Wisconsin. Alonzo Pawling worked in Milwaukee as a wood patternmaker and he and Maurice Weiss established the Milwaukee Tool and Pattern Shop in 1883. The following year the partnership was dissolved and Henry Harnischfeger joined Alonzo Pawling, and Pawling & Harnischfeger was founded to build knitting machines. The company soon became a job manufacturer for the products or components of others. From about 1887 to 1892 P & H and H. A. Shaw formed a separate company to manufacture overhead electric cranes, and later P & H made its own overhead cranes, also manufacturing their electric motors and controls. During the first decade of this century P & H built its first, but

unsuccessful, excavating machine to the designs of Mr Hettelsater, a former employee of Allis-Chalmers. In 1911 Alonzo Pawling relinquished his interests in P & H and soon the company became the Harnischfeger Corporation but retained its original trademark. Around 1913 a bucket chain trencher was produced and shortly afterwards L. Wehner was employed, from Bucyrus, to develop P & H excavating machinery. Buckeye-type bucket wheel trenchers were built and a boom type ditcher was superseded by a crawler mounted dragline which became the 210 produced from 1914 to 1925. A half track 0.38 m³ (½ cu yd) 205 model followed in 1919, which became the full track 206 in 1920 when full development and promotion of excavators commenced. Many models of excavator were produced in the coming years, culminating in the development of a series of very large electric mining shovels which was initiated by the erection of the first 19 m³ (25 cu yd) 2800 model in 1968. P & H, which had produced its first electric shovel in 1933, still uses its own electrical equipment, now in excavators, as it had done on its first overhead cranes.

In 1964 P & H acquired the backhoe recently developed by the Cabot Corporation's Machinery Division at Pampa, Texas; a line of hydraulic machines followed including, from 1970 to 1974, O & K models made under licence.

P & H has always made other products – now including materials handling equipment, hoists and electric motors for outside users, but primarily it is a producer of cranes and excavators. The current range consists of nine basic models of dragline, the largest being the 2355 introduced in 1981 with bucket capacities up to 13.8 m³ (18 cu yd); the 1200 hydraulic backhoe or shovel weighing up to 173 t; and six basic models of electric mining shovels up to the 45.9 m³ (60 cu yd) 5700. With the exception of two models of dragline and the 1400/DE shovel, all current P & H models are built in the USA at either West Milwaukee, Wisconsin, or Cedar Rapids, Iowa; in 1983 the Escanaba, Michigan, factory closed.

As early as 1935 P & H gave a licence to Newton Chambers, and NCH 100, 125 and 150 models were made in the UK during the period up to 1947. Kobe Steel have produced P & H

The first P & H 2355, largest of crawler draglines, has worked in Alabama since 1981.

excavators since 1955 in Japan. The 1200 was developed and originally built at Dortmund, West Germany, by Harnischfeger GmbH – a partnership created in 1974 between Harnischfeger and Rheinstahl AG which had made P & H crawler machines since the 1950s. P & H excavators are also made in Iran, India and Brazil; components are made at P & H factories in Australia and South Africa and by licensees in Canada and South Africa; and in 1984 major parts of two 2300s, destined for Zambia, were built by subcontractors in the UK.

UNIT (HYDRA UNIT) Unit Crane and Shovel Corporation

William Ford, the brother of Henry, in 1925 formed the Wilford Shovel Co at Highland Park,

Michigan, to manufacture a 270° slew, 0.19 m³ ($\frac{1}{4}$ cu yd) shovel. The name was soon changed to the Universal Power Shovel Company and in 1928 the Unit Corporation of America bought the firm and began manufacturing excavators at the Unit Drop Forge factory in the Milwaukee suburb of West Allis, Wisconsin. Before long the excavator business was sold again, to a group from Unit Drop Forge, and the name 'Unit' began to be used on excavators. The early limited slew shovels of 0.19 m³ ($\frac{1}{4}$ cu yd) and 0.57 m³ ($\frac{3}{4}$ cu yd) capacity were mounted on truck chassis or Buckeye crawlers; they used Fordson farm tractor engines and components, and were

A Soviet EO-5123 showing its versatility.

convertible to backhoe or crane. In 1940 the present owners bought the company and it was for a time known as the Universal Unit Power Shovel Corporation. Another excavator manufacturer, Bay City – established in 1913, was amalgamated with Unit, in 1961, and its excavators soon phased out. Prior to 1965 Unit produced a range of crawler and wheeled cable machines such as the 1020; in 1965 the H-201C hydraulic backhoe was introduced. Since 1967 Unit excavators have been made at New Berlin, Wisconsin, and the range in the late 1970s included four models of hydraulic backhoes from the 21 t H202C Series II to the H-47 weighing up to 47 t. Unit were also manufacturing offshore cranes.

USSR

Soviet Government
Excavators have been used in this country since Otis shovels were imported in the 1840s; and in the early years of this century Bucyrus machines were made under licence in Leningrad with, around the same time, the first Marion unit arriving destined for work on the Murmansk Railway. Despite this, of the hundred and fifty excavators in the USSR in 1922, only twenty-five were Soviet-made. In 1931 the first steam powered universal excavator was manufactured at the Kovrov factory at Kirov, north west of

Moscow, and series production subsequently began. Now there are at least eleven factories producing excavators in the USSR.
KOVROV EXCAVATOR FACTORY
This now produces hydraulic excavators with bucket capacities of 0.5–1 m³ ($\frac{5}{8}$–$1\frac{1}{4}$ cu yd), including the EO-412A.
RABOCHI EXCAVATOR FACTORY
Situated at Kostroma, north east of Moscow, this produces machines up to 1 m³ ($1\frac{1}{4}$ cu yd) capacity.
KALININ EXCAVATOR FACTORY
The excavator factory in the town of Kalinin, north-west of Moscow, manufactures hydraulic excavators up to 0.6 m³ ($\frac{3}{4}$ cu yd) capacity.
KOMINTERN EXCAVATOR FACTORY
Voronezh, the location of the Komintern factory, is south of Moscow. This was established in the 1920s and produces cable excavators – the E-1252B and E-2503, and hydraulic excavators – the EO-5122 introduced in 1975 and presumably the EO-5123, and the EO-6122 introduced in about 1978.
LENINGRAD EXCAVATOR FACTORY
Wheeled and crawler hydraulic machines up to 3 m³ (4 cu yd) are manufactured here.
IZHORSK EXCAVATOR FACTORY
The Izhorsk factory is also in Leningrad and produces larger excavators, between about

3.2–12.5 m³ ($4\frac{1}{8}$–$16\frac{3}{8}$ cu yd), including the cable EKG-4U and EKG-12.5.

KRASNY EXCAVATOR FACTORY

Three excavator manufacturing factories are situated in the Ukraine; this one is at Kiev to the west. Hydraulic backhoes are made here – the wheeled EO-4321, the crawler EO-4123 and EO-5015A. Backhoe loaders are also produced.

NOVOKRAMATORSK ENGINEERING WORKS (NKMZ)

Also in the Ukraine is the NKMZ factory established at Kramatorsk in 1948. This manufactures large cable shovels, generally up to about 25 m³ (33 cu yd) together with walking draglines including the ESh-6/45M and ESh-10/70A. Other products include bucket wheel excavators, mine hoists, ore crushers, rolling mill equipment and hydraulic presses.

DONETSK EXCAVATOR FACTORY

The third Ukranian excavator factory was established at Donetsk in the late 1970s. Excavators of 0.65 m³ ($\frac{7}{8}$ cu yd) capacity are produced.

URALMASHAVOD HEAVY ENGINEERING WORKS (UZTM)

The UZTM factory at Sverdlovsk in the Ural Mountains was founded in 1933 and included in its initial output was a 3 m³ (4 cu yd) capacity electric powered excavator. Amongst its excavator products are the EKG-5A and EKG-20 shovels and other cable shovels, generally up to about 25 m³ (33 cu yd); the ESh-15/90A and other basic walking draglines of 25, 40, 65 and 80 m³ (33, 52, 85 and 105 cu yd); and since the late 1970s it has introduced its first hydraulic excavators, the EG-12 and EG-20. The factory also makes a wide range of other heavy engineering products.

KRASNOYARSK EXCAVATOR FACTORY

At Krasnoyarsk in Siberia is what appears to be the only factory manufacturing excavators outside European Russia. Large cable shovels are produced, generally up to about 20 m³ (26 cu yd).

Soviet excavators are made under licence in China and India.

WEST GERMANY

ATLAS   H Weyhausen KG

Founded in 1919 by Heinrich Weyhausen, the company produced agricultural machinery and its first fully hydraulic AB1500 excavator emerged from the Delmenhorst works in 1954. The main current line of eight basic models of hydraulic machines are developments of a new generation of Atlas excavators introduced in 1968; weights range from 7.3 to 53 t. In 1982 two models of mini backhoes were added, built by Iwafuji in Japan and modified by Atlas. The company's headquarters is at Delmenhorst where prototypes are made; excavators are assembled at the Ganderkesee factory with part-manufacture being undertaken at the Vechta and Löningen factories. From 1966 to 1977 Atlas excavators were made by Kubota in Japan; machines have been assembled in Tehran, Iran; and for nearly three years in the late 1970s five models of Atlas excavators were marketed in North America by Deere. Atlas also produces truck and agricultural loading cranes, truck tail lifts and skip carriers, wheel loading shovels and concrete placing booms.

BAVARIA   Gebr Hofmann Maschinenfabrik und Eisengießerei

The Bavaria range of three hydraulically driven crawler draglines, from Eibelstadt-Würzburg, weigh overall up to 14 t and have bucket capacities up to 0.8 m³ (1 cu yd). Hofmann products include a range of mobile cranes and agricultural machinery.

DEMAG   Mannesmann Demag AG
Mannesmann AG

Wekstaette Harkort & Co, Duisburger Machinen AG and Benrather Maschinenfabrik AG, founded in 1819, 1862 and 1891 respectively, were manufacturers of industrial machinery and suppliers to the steelmaking and mining industries of the Ruhr Valley. They merged in 1910 to form Deutsche Maschinenfabrik AG (Demag). In the early 1920s the company's first excavator, a rail mounted GII steam shovel, was made; crawler mounted machines followed in 1928 and in 1936 diesel power was introduced. Demag's largest cable excavator was the 6.7 m³ ($8\frac{7}{8}$ cu yd) electric U35 made between 1942 and 1944. From 1965 the company's cable machines were phased out, except the B406 and B410 which were available as draglines or grabs – the B410-LCB until quite recently. In 1954 Demag launched its first hydraulic excavator, the crawler B504 with a standard 0.4 m³ ($\frac{1}{2}$ cu yd) bucket and developed from the B304 cable machine. Until

By the end of 1983 over sixty Demag H241s had been built in the first five years of production.

1939 Demag excavators were made at Duisburg; since then they have been produced at Düsseldorf-Benrath. The present H range, evolved from the H41 first produced in 1968, consists of six models, all available as shovel or backhoe. These range from the 40 t H40 introduced in 1983 to the H241 introduced in 1978 and now weighing some 270 t. Demag's subsidiary in Brazil manufactures three models of hydraulic excavator and production started in China in 1984 of the H55 and H85 under an agreement reached in 1983. Demag has been involved with Richier excavators in the past, and from 1977 to 1980 Hymac marketed Demag excavators in the UK. In 1974 Demag joined the Mannesmann AG group and now makes machines such as cranes, road finishers, bucket wheel excavators, conveyors and stackers.

EDER   Eder Hydraulikbagger GmbH
Eder GmbH & Co KG

The Eder company of Mainburg, 80 km (50 miles) north of Munich in Lower Bavaria, was established by Franz Eder in 1946 who started in business as a blacksmith. Draglines were produced from about 1953 to 1963, then hydraulic excavators were introduced. Now Eder manufactures hydraulic cranes and excavators only and in the late 1970s opened an additional factory at Eging am See, also in Bavaria, but near Passau adjoining the Austrian border. Eder has specialised in making excavators to be marketed by other companies; the Hanomag M40 (1970–74) was the Eder M40, the Hanomag/ Massey-Ferguson 250 (1974–80) was the Eder M-400 and the Hanomag 550CLC (1978–80) was the Eder R835LC; between 1980 and 1981 both the International Harvester 620W and the Brøyt X12 were Eder M815s; and from 1984 five crawler and four wheeled versions of the four basic Eder models are being marketed by Caterpillar. All Eder's current '800' series, initiated with the 835 in 1978 and the most recent, the R805 introduced in 1984, are available with wheel or crawler mountings; weights range from 11 to 23 t.

EXCAVATORS

EWK (COMBI CRAFT) Eisenwerke
Kaiserslautern Göppner GmbH

In 1864 at Kaiserslautern in the Rheinland-Pfalz
area, the Eisenwerke company was founded;
early products included castings for such things
as heating systems. The company later made
cranes and steel for bridges; its products now
include military and tunnelling equipment. Two
models of telescoping boom excavators are made
at Kaiserslautern: the 19 t PL820R (Combi
Craft) which is crawler mounted, and the
PL820MF truck mounted version.

FAUN Faun Frisch Baumaschinen GmbH
Faun-Werke Zentralverwaltung

Faun's range of four hydraulic backhoes result
from the products of Karl Mengele & Sohne of
Gunzburg-Donau; a company founded in 1872
which had made excavators since 1957.
Mengele's excavator division was acquired by
Faun in 1978. Four models are currently produced, three wheeled FM models range from 11
to 19 t and the crawler FR1035LC weighs up to
19 t. All excavators are made at the Neunkirchen
factory near Lauf a d Pegnitz, opened in 1948.
The predecessor company of Faun dates back to
1845 and the present company's name derives
from a merger in 1918 of factories in Ansbach and
Nürnberg to form Fahrzeugwerke Ansbach Und
Nürnberg AG (FAUN). The Faun Frisch division was created after the acquisition in 1977 of a
majority interest in Frisch GmbH. Faun produces large specialised commercial vehicles, construction equipment and municipal vehicles.

FUCHS Johannes Fuchs KG

Since it was established in 1888, Fuchs has been a
family business. It has made excavators since
1955 and has factories at Ditzingen near Stutt-
gart and Bad Schönborn near Karlsruhe. The
first series of machines in the present range
consists of two basic models of crawler hydraulic
backhoes both also available on wheeled mountings, together with another wheeled model; these
were introduced in either 1975 or 1978 and weigh
from 12 t to the 20 t 714R. The second series, of
hydraulically driven draglines, comprises two
basic crawler models also available on wheeled
mountings, together with a crawler and two
wheeled models; these were introduced between
1977 and 1980 and weights range from around
9.9 t to the 125R weighing up to 34 t. The earlier

Fuchs 703R was made by Whitlock in the UK as
its 50R around 1970, and production of the 702M
by Müller of Rio de Janeiro, Brazil, has recently
ceased. Fuchs also manufactures aggregate handling machinery and agricultural equipment.

LIEBHERR Liebherr-Hydraulikbagger
GmbH

Liebherr-Holding GmbH

Hans Liebherr, born in 1915, served his apprenticeship as a bricklayer in his stepfather's building
firm in the village of Kirchdorf on the River Iller,
Upper Swabia. In 1930 he qualified as a master
builder and took over the family business. While
still operating this business in 1949 he devised a
tower crane which was assembled in a smithy in
Kirchdorf and tested by his own firm. Liebherr
was then established as a manufacturing concern,
gear cutting machines were also soon produced
and in 1954 the first Liebherr fully hydraulic
excavator, the three wheeled 7.5 t L300 with a
0.9 m³ (1⅛ cu yd) bucket was launched from
Kirchdorf. Hans Liebherr had been prompted to
produce his own hydraulic machine when he
questioned the high weight/bucket size ratio of a
cable excavator hired for a small job being
undertaken by his building firm. Liebherr now
has a range of eleven basic crawler mounted (R)
hydraulic excavators: nine are '2' series machines
introduced from 1980 onwards and five are
available as wheeled (A) versions. Weights range
from the 12 t A900 to the 164 t R991 introduced
in 1977. Only six small models are made at the
original Kirchdorf factory – the A and R 900, A
and R 902, A912 and A922; most models are
made at Colmar, France and three models at
Newport News, Virginia. Liebherr's Nenzing
factory in Austria produces three models of
hydraulically driven draglines, the first introduced in 1981.

From 1972 to 1979 Kobe Steel had a licence to
manufacture two models of Liebherr excavators
in Japan and these were made between 1973 and
1978; in the late 1970s R981s were made in
Romania; in 1982 production of the R942 started
at Shanghai, China, and it is intended to make the
R991 in Brazil. The Liebherr group of companies produces a variety of construction machines, specialised cranes, equipment for the
aircraft industry, refrigerators and machine tools
– as well as operating hotels. Since 1976 Liebherr

has functioned as two autonomous groups, one based in Germany, the other in Switzerland; all capital of both companies is held by the five children of Hans Liebherr.

O & K   Orenstein & Koppel AG

O & K's roots can be traced back to 1876 when Benno Orenstein and Arthur Koppel established Orenstein & Koppel OHG at Berlin Schlachter-see, to manufacture foundry equipment. Although the partnership was dissolved in 1885 and two separate companies formed, co-operation was re-established in 1905 and a merger took place four years later. In 1911 O & K acquired a majority shareholding in Lübecker Maschinenbau Gesellschaft (LMG) and complete incorporation occurred in 1950. Narrow gauge railway equipment had soon become a speciality of the early companies and before long heavy locomotives and rolling stock were added; later railway construction world-wide was undertaken. Both predecessor O & K companies

and LMG had commenced production of bucket chain excavators in the last decades of the nineteenth century. O & K produced its first rail mounted steam excavator in 1908 at the Berlin-Spandau factory established in 1900. The association with LMG from 1911 opened the way for production of larger excavators and dredges. In 1922 the Berlin factory produced the first crawler shovel, and in 1926 the universal Type 4 excavator was launched. By the 1950s cable machines, such as the 'L' series of universal crawler excavators with capacities up to 2 m³ (2⅝ cu yd) were in production; although O & K had already made larger models. In 1961 the first hydraulic excavator, the RH5, appeared and series production commenced at the Berlin and Dortmund (opened in 1894) factories. Currently the smaller models are made at Berlin and

Two O & K backhoes from a range developed since the RH5 went into series production in 1961.

medium and large sizes at Dortmund. O & K purchased a factory at Dundas, Ontario, in 1974 and for a limited time excavators, including the RH75, were made there. From 1964 to 1979 JSW made O & K excavators in Japan, and P & H manufactured O & K excavators in the USA from 1970 to 1974. From 1984 O & K are marketing in the UK two Macmoter models: the 4.6 t 32B and 10 t 58B. Twelve basic models are currently in production; eight are available only as crawler machines, three either on crawler or wheeled undercarriages, and one only as a wheeled machine. The largest is the RH300 introduced in 1979 which, at 485 t, is easily the largest production model of hydraulic excavator in the world. In 1982 'C' versions of some models appeared with improved front end geometry. The most recent model is the 184 t RH120 introduced in 1983. A prototype 'Futura' based on the RH9, was also exhibited in 1983. O & K products include wheel loaders, bucket wheel excavators, fork lift trucks, escalators, railway rolling stock, ships and dredges.

SCHAEFF   Karl Schaeff GmbH & Co
Engineering workshops were established by Karl Schaeff at Langenburg, Swabia, in 1937; later the company began to produce attachments for construction equipment and tunnelling machinery. As well as excavators, at the factories at Langenburg, Rothenburg ob der Tauber and Crailsheim, products include wheel loaders, fork lift trucks and backhoe loaders. The present range of excavators comprises five crawler machines from the 2.2 t HR3 to the 6.8 t HR26D – the three smallest are Nissan mini models, one modified, and were introduced in 1983; together with two wheeled machines, the 4.8 t HML15A and 6.9 t HML30D. Schaeff excavators are fitted with a 'Knickmatic' device which allows digging parallel to, and alongside, the crawlers. In addition, there is a climbing-type excavator, the HS40A and an HRS2B special backhoe for narrow spaces; also Schaeff excavators are mounted on Mercedes-Benz Unimog truck chassis.

SENNEBOGEN   Maschinenfabrik Sennebogen GmbH
Erich Sennebogen and a partner founded Sennebogen & Six in 1952. The company initially produced agricultural machinery before diversifying to manufacture small cranes and grabs

A 55 t Weserhütte SW190 excavator with an earth auger.

for the agriculture and building industries. Sennebogen's present range of hydraulic backhoes and hydraulically driven draglines, developed from models first created in the early 1960s, are produced at Straubing-Donau. Four hydraulic backhoes with weights from 11 to 24 t are available in crawler or wheeled versions; three of the wheeled versions – the 212, 214 and 216 – are marketed by the West German firm of Zeppelin. The four draglines, again in crawler or wheeled versions, range from 17 to 30 t. Sennebogen base machines are also available as cranes on a variety of mountings.

WESERHÜTTE   PHB Weserhütte AG
PWH Group – Otto Wolff AG and Hoesch AG
Weserhütte was established in 1844 and its first excavator, a rail mounted steam shovel, was developed in 1904 but not introduced until 1908 as the L8. Other L series models followed; the L5, 11 and 18. LR crawler mounted and diesel powered models were next; then the W300-W1600, W2-W24 and eventually the

W40-W270/320 series initiated in 1964. At Bad Oeynhausen, Weserhütte now produce two ranges of excavators. Present models in the W range of three basic crawler mechanical machines – 100/120, 160/180 and 270/320 – were introduced between 1967 and 1976 and are available as shovels up to 3 m³ (4 cu yd) or draglines up to 3.8 m³ (5 cu yd). Five SW crawler hydraulically driven draglines – 140, 190, 320, 530 and 760 – were introduced between 1978 and 1982 with overall bucket capacities of 0.8–8.4 m³ (1–11 cu yd); the SW1220 of 12.2 m³ (16 cu yd) was added in 1984. In 1966 Weserhütte began making Hydrowolff hydraulic excavators, a series including the HW70 and HW130, but production ceased in the late 1970s.

Weserhütte AG was part of the Otto Wolff Group when it was joined by Pohlig-Heckel-Bleichert AG in 1980, and PHB Weserhütte AG was formed as a division of the PWH Group. PWH specialises in materials handling and processing equipment such as bucket wheel and chain excavators and spreaders, as well as producing equipment for the nuclear power and packaging industries. Since 1983 there has been an ownership link with O & K through Hoesch AG.

YUGOSLAVIA
POCLAIN   Duro Daković
Production of the 125CK was expected to commence at Slavonski Brod during 1983.

CHAPTER 10

Scale Models of Excavators

Man has an apparent innate desire to make and collect small-scale images of things around him. The fascination engendered by such objects, which exists universally in young and old, is clearly exploited to the full by toy and model manufacturers. The engineering industry is no less aware of the interest in, and value of, scale models: they help designers to plan and develop ideas; are used in promotion exercises to assist potential customers in visualising and understanding the working of finished products – especially if these are complex or too large to demonstrate; and are employed in training for those making or operating machines.

The excavator manufacturing industry has for a long time used scale models for all these purposes. In 1878, for example, Priestman showed a model of a grab crane at the Paris Exhibition where the Prince of Wales, later to become Edward VII, intimated that it would make a nice plaything for his children; but the hint was not acted upon! The following year the same model was taken on the long voyage to Sydney, Australia, to promote grab cranes. Bucyrus-Erie's 120-B full slew crawler shovel appeared in 1925 and was a direct competitor to railroad shovels. Its advantages were well demonstrated soon after its introduction by B-E at exhibitions with a display of working models of a 120-B and a railroad shovel loading railway wagons.

Scale models as aids to design have included Hitachi using one of a UH20 hydraulic shovel at meetings of the development team; the conceptual basis of Marion's 204-M Superfront developed to prototype stage through a series of wooden working models; and B-E had a small-scale model of the 3270-W walking dragline to illustrate in detail its internal structure, and used a 'test bed' quarter-scale model of the 550-HS hydraulic mining shovel.

Part of the training programme of apprentices often includes building scale models; a quarter-scale model of a Hymac 580 backhoe was built by apprentices at Hymac's Rhymney factory. At the apprentice training school of the UZTM factory in the USSR, large scale models of walking draglines are part of the training equipment. For driver instruction Ruston-Bucyrus uses a working 1:20 scale model of a 110-RB shovel controlled from a full sized facsimile of the cab.

The most extensive, and most evident, use of scale models in the excavator industry is in promotional exercises. Two types of scale models are used: specially built 'one-off' representations, sometimes quite large and with powered working parts, used for exhibitions or presentations to prominent people; and small commercially mass-produced units for widespread distribution to customers and others. Brøyt have a large and detailed 'cut away' model of the X50 for exhibition purposes. Amongst significant presentations was that in the USA to President Eisenhower in 1960. Officers of the National Coal Association gave the President a working model of a Bucyrus-Erie 3850-B stripping shovel, at a scale of about 1:200 and made of wood and aluminium, as a present for his grandson, David. President Eisenhower later wrote, 'I am still contemplating with awe the statistics you gave me concerning the shovel'. In 1983, while the UK's Duke of Edinburgh was on a tour of Australia, he was presented with a static scale model of a Marion 8050 walking dragline, being built for work at coal mines in Queensland operated by Capricorn Coal Management Pty Ltd.

Detailed 'one-off' scale models of excavators can be built by professional model-making firms, by professional or amateur model-makers employed by companies which manufacture the real machines, or by outside amateurs pursuing their

hobby. In the UK Bassett-Lowke (SM) Ltd, one of the earliest makers of model railways and now a specialist model-making company, has produced a scale model of a 38-RB dragline for Ruston Bucyrus. Similarly, Severn-Lamb Ltd, the first English model-makers to produce an exhibit for the Smithsonian Institute in Washington DC, has built scale models of the 'Cub' for Priestman, a 1 : 10 808 backhoe for JCB, and two 1 : 75 W2000 walking draglines for Ransomes & Rapier.

Scale models made by employees of excavator manufacturers have included those of a P & H 5700LR shovel that took 329 hours to make at Milwaukee, Wisconsin; and a working 1 : 12 model of a Smith 21 shovel (a 0.01 t machine!) made at Rodley, West Yorkshire. One of the most impressive working scale models produced by an amateur model-maker not employed by an excavator manufacturer is that of a Bucyrus-Monighan 10-W walking dragline. This 1 : 32

NZG 1 : 50 Demag H185 loads Conrad 1 : 50 Wabco 170 truck, with a Manitowoc 6400 in the background.

In 1983 the Duke of Edinburgh was presented with a scale model of a Marion 8050 walking dragline when he was on a tour of Australia.

President Eisenhower was presented with a model, at a scale of about 1 : 200, of a B-E 3850-B stripping shovel at the White House in 1960.

The only commercially-made scale model with cast metal crawlers currently available, Solido's 1 : 55 International/Yumbo 3984; Tekno's Åkerman 700 backacter in the background.

machine was made by Walter Bennett in Pennsylvania between 1967 and 1981. Patterns were generated from aluminium and/or plexiglass and cast in brass. The machinery house is constructed of brass plates; even rivets were soldered to the boom to simulate the actual joint construction. All doors and windows can be opened, and the driver's cab is an exact replica of the original. Each walking cam mechanism consists of five separate castings and the bucket has replacement teeth and ball bearings in the sheaves. Three DC electric motors provide power and when operating the model's speed and acceleration are nearly identical to the original machine shipped in 1934.

The first group of mass-produced models of excavators comprises those welcome 'construction', rather than more common 'destruction', ones made as toys; in which case although based on actual excavators they may not be exactly true to scale and be modified for use by children to be safe or so that some simple working mechanism can be incorporated. A second group covers detailed scale models essentially produced as collectors' items for display; however, many models cannot readily be classified into one or

A fine example of the model engineer's craft is this 1:12 working model of a Smith 21 chain crowd shovel built at Rodley, West Yorkshire, by Smith employees.

other of the two groups. At least two hundred generally available different scale models of excavators have been produced in countries including Austria, Denmark, East and West Germany, France, Italy, Japan, Spain, the Netherlands, UK and USA. Predominant amongst these manufacturers are the West German firms of Nürnberger Zinkdruckguss-Modelle (NZG) of Nürnberg, which has specialised in zinc die-cast models of construction equipment since 1968, and Conrad GmbH of Kalchreuth-Röckenhof which incorporates earlier Strenco and Gescha trade names; together with Yonezawa Toys Co Ltd of Tokyo, Japan, which makes Diapet models. These three companies alone account for a total of around a hundred of the versions made and between them have marketed representations of Poclain, Atlas, Demag, Eder, Faun Frisch, Fuchs, Liebherr, O & K, Schaeff,

Sennebogen, Weserhütte, Massey-Ferguson, Hitachi, IHI, Kobelco, Komatsu-Bucyrus, Komatsu, Kubota, Mitsubishi, Sumitomo, Yanmar, Brøyt, Åkerman, Hymac, JCB, Priestman, R-B, Whitlock, B-E, Case Drott, Caterpillar, FMC Link-Belt, Lima and P & H excavators.

Most of the types of cable and hydraulic excavators are illustrated by mass-produced scale models – except draglines and stripping shovels; crawler and wheel-mounted machines are covered, and Rio Grande Models, California, market 1:87 kits of a Marion 40 and an American Ditcher railroad shovel. There is, however, only one model of an electric mining shovel, that at 1:87 by Gescha of a P & H 2800; the only cable backhoes are the Åkerman 610, 700 and 752 models made by Tekno at factories in Denmark and the Netherlands, the USA 1:87 kit by Scenic Details of an Insley K-12 and the Austrian/USA 1:87 by Umex/AHM of a Unit 271C; and two skimmers, the UK Moko of a Rapier 410 and USA Structo probably of a Keystone half-track.

Although scales range from Davco's 1:148 kit of a Priestman Mustang 220 backhoe, to Triang's 1:18 of a Rapier 423 shovel – both made in the UK – the majority of models are at a scale of 1:50. Scales often relate to those of model railways, including 0 (1:43), 00 (1:76), H0 (1:87) and N (1:148) gauges. Tinplate was used for early examples such as Nylint's Michigan T24 grab and C24 shovel, Doepke's Unit 357 grab and Buddy L's W & S Gradall – all made in the USA.

More recently plastic has been used for such models as the series of fifteen Poclain excavators by the French Bourbon company. The majority of models, though, are made of die-cast metal and are often composite creations including, for example, rubber or plastic crawlers (only the French Solido's Yumbo-International 3984 backhoe is currently available with cast metal crawlers).

Most scale models have parts that move but some are also designed to be operated by levers and handles, often incorporating ingenious mechanisms. The 1:24 Lorain TL-25 shovel made in the USA by Reuhl is assembled from eighty-one components, fully reeved, and all the excavator's digging functions can be performed by operating handles. Similarly, two UK models, the Dinky 38-RB and Corgi Priestman Cub V

Dinky 38-RB shovel improved by adding the proper livery; Whalley, Lancashire, is the background.

shovels are worked by turning knobs – drivers' instructions are enclosed!

Standard mass-produced scale models can be made more authentic or altered by adding different front-end equipment. The Dinky 38-RB can be greatly improved by repainting in R-B colours and adding decals and livery markings, and the NZG Fuchs 118R lattice boom crane can easily be converted into a dragline by adding a fairlead and bucket.

In West Yorkshire Ron Smyth has created finely detailed representations of Atlas AB1102D, AB1202D, AB1302DK, AB1602D (with high rise cab and cuff link-derived magnet) and AB1902D machines from Conrad AB1302D and NZG AB2002D models.

A limited number of old excavators are pre-served. For example, a 1924 Ruston No 6 in Hertfordshire, a 1926 Keystone No 4 in Pennsylvania, a 1941 10-B in Alberta, a 1941 Åkerman 300 in Norway and a 1948 Rapier 4½0 in the Netherlands are in private hands. UK museum collections include three steam shovels in working order – a 1909 Ruston Proctor at Lincoln, Lincolnshire, a 1931 Ruston No 25 at Beamish, County Durham, and a 1935 52-B at Leicester, Leicestershire; and in North American museums are Erie-Bs dating from the 1920s at Atlin, British Columbia and a Marion 21 of about 1922 vintage at Bellows Falls, Vermont. Due to their size and often inaccessible working locations, though, relatively few old excavators remain intact; the creation, preservation and display of scale models of such fascinating machines is one way of maintaining recognition of the heritage of this vital section of the world-wide engineering industry.